D0952197

TECHNOPOLES OF THE WORLD

Cities and regions are being profoundly modified in their structure, and conditioned in their growth dynamics by the interplay of three major historical processes: technological revolution, the formation of a global economy, and the emergence of an informational form of economic production and management.

Technopoles, planned centres for the promotion of high-technology industry, are the reality of these fundamental transformations, redefining the conditions and processes of local and regional development. Generating the basic materials of the informational economy, technopoles constitute the mines and foundries of the informational age, yet have received less attention than their counterparts of the industrial economy.

Technopoles is the first comprehensive survey of planned development in all its manifestations: science parks, science cities, national technopolis and technobelt programs. Case studies, ranging from Silicon Valley to Siberia, from the M4 Corridor to Taiwan and Korea, relate how technopoles begin, how they take shape, what each is striving to achieve and how well it is succeeding.

Embracing a host of disparate concepts, and a few myths, *Technopoles* distills the lessons from the successes and the failures, offering guidelines for national, regional and local planners and developers worldwide.

Manuel Castells is Professor of City and Regional Planning at the University of California, Berkeley, and Professor of Sociology at the Universidad Autonoma, Madrid. Peter Hall is Professor of Planning at University College London, and Professor Emeritus of City and Regional Planning at the University of California, Berkeley.

TECHNOPOLES OF THE WORLD

The making of twenty-first-century industrial complexes

Manuel Castells and Peter Hall

London and New York

First published 1994
by Routledge
11 New Fetter Lane, London EC4P 4EE

Simultaneously published in the USA and Canada
by Routledge
29 West 35th Street, New York, NY 10001

Reprinted 1996

Routledge is an International Thomson Publishing company

Typeset in Garamond by
J&L Composition Ltd, Filey, North Yorkshire
Printed and bound in Great Britain by
Biddles Ltd, Guildford and King's Lynn

British Library Cataloguing in Publication Data
A catalogue record for this book is available from the British Library

Library of Congress Cataloguing in Publication Data
A catalogue record for this book is available from the Library of Congress

ISBN 0-415-10014-3 (hbk)
ISBN 0-415-10015-1 (pbk)

CONTENTS

FIGURES

TABLES

ACKNOWLEDGEMENTS

Our first thanks go to the Sociedad Estatal Expo '92 in Seville, Spain, for providing the financial support for the research on which this book is based. This support – specifically to analyze foreign experience as a basis for the construction of a technology center on the Expo '92 site in Seville (the Cartuja '93 project, described in Chapter 8) – funded our field work in Spain, France, Germany, the United Kingdom, Russia, Japan, Taiwan, and the Republic of Korea, and provided us with research assistance in analyzing our data. We are particularly grateful to Mr Ignacio Quintana, director of the Cultural Division of Expo '92, and to Mr Antonio Palaez, first coordinator of the Cartuja '93 project, for their invaluable support; without it, the study presented here would never have been possible.

We also wish to acknowledge financial support from the Committee on Research at the University of California at Berkeley, which made possible the production of the illustrations in this book, and from the Instituto Universitario de Sociologia de Nuevas Tecnologias at the Universidad Autonoma de Madrid. In addition, the Korea Research Institute for Human Settlements (KRIHS) provided institutional support for our work in Korea; and the Institute of Economics and Industrial Organization of the Russian Academy of Sciences, Siberian Branch, in Novosibirsk, did the same for our Russian investigations. In Berkeley, we were most ably assisted in our research by two outstanding graduate assistants of the Department of City and Regional Planning: You-tien Hsing and Hiro Izushi.

During our field work we were welcomed and assisted by officials and technical collaborators in many countries, too numerous to name individually; we should like to record our sincere thanks to all of them for the time and trouble they took to answer our questions. Specifically, we were afforded cooperation by colleagues who performed an invaluable role in smoothing our logistical and intellectual path through the interviews: Professor Joo-Chul Kim, in the Republic of Korea; Ms Natalia Baranova, in Novosibirsk; Professors Takashi Machimura, Dai Nakagawa, Tsunekazu Toda, and Mamoru Taniguchi, and Messrs Manuel Sanchez and Pedro de Mingo, in Japan; Dr Semra Téber and Professor Guy Benveniste, in France; Dr

Carmen Hass-Klau, in Germany; the principals of Segal Quince Wicksteed, in Cambridge; and Dr Clara Garcia and Dr Isabel Ramos in Seville.

In Berkeley, Martha Conway and Jane Sterainger performed an impeccable job in developing the software that produced the maps.

Finally, we acknowledge with thanks the service performed by a number of academic colleagues in different countries who commented on the manuscript in draft: Professors Jean-Paul de Gaudemar in Strasbourg, John Goddard in Newcastle upon Tyne, and AnnaLee Saxenian in Berkeley. Stephen Hamnett deserves special mention for providing exceptionally speedy last-minute corrections to the Australian MFP story.

Manuel Castells
Peter Hall
Madrid and London, January 1993

1

TECHNOPOLES: MINES AND FOUNDRIES OF THE INFORMATIONAL ECONOMY

There is an image of the nineteenth-century industrial economy, familiar from a hundred history textbooks: the coal mine and its neighboring iron foundry, belching forth black smoke into the sky, and illuminating the night heavens with its lurid red glare. There is a corresponding image for the new economy that has taken its place in the last years of the twentieth century, but it is only just imprinting itself on our consciousness. It consists of a series of low, discreet buildings, usually displaying a certain air of quiet good taste, and set amidst impeccable landscaping in that standard real-estate cliché, a campus-like atmosphere.

Scenes like these are now legion on the periphery of virtually every dynamic urban area in the world. They appear so physically similar – outside Cambridge, England or Cambridge, Massachusetts; Mountain View, California or Munich, Germany – that the hapless traveler, dropped by parachute, would hardly guess the identity of the country, let alone the city. The developments they represent go under a bewildering variety of names, which invariably permute a few key elements like Techno, Science, 21st Century, Park, Plaza, Polis, and -topia. In France, where there are certainly as many of them as anywhere else, they go under a generic name, *technopole*. It is so evocative that in this book we have decided to appropriate it for the English language.[1]

Generally, technopoles are planned developments. Some are pure private sector real-estate investments, and these happen to be among the most numerous but least interesting. A significant number, however, have resulted from various kinds of cooperation or partnership between the public and private sectors. They are promoted by central or regional or local governments, often in association with universities, together with the private companies that occupy the resulting spaces. And these technopoles, the more interesting ones, are invariably more than just plots to rent. They also contain significant institutions of a quasi-public or nonprofit type, such as universities or research institutes, which are specifically implanted there in order to help in the generation of new information. For this is the function of the technopole: it is to generate the basic materials of the informational economy.

Quite a number of the people in the buildings of these new technopoles do not usually make anything, though somewhere else, often not many miles away, in rather similar buildings – sometimes of slightly less elegance – other people are making the things they invent here. The things may be computers or VCRs or CD players, word processing or spreadsheet software, artificial systems, high-technology ceramics, genetically engineered drugs, or a thousand other new products. What the things have in common is that they embody information that has essentially been created here. These high-technology products – hardware and software, bulky products and almost immaterial ones – are the products and symbols of a new economy, the informational economy. The information they embody has been created in technopoles, and invariably the embodiment of the information into the products also occurs in technopoles, which thus constitute the mines and foundries of the informational age.

The informational economy has been less noticed than the industrial economy it is replacing, and technopoles have had less attention from academic analysts than factories and mills. Though there are books about science parks in individual countries, and isolated articles about bigger experiments like the Japanese technopolis program, there is no account that tries to bring together, in one descriptive-analytic account, the most important ventures in constructing technopoles worldwide.

This book has been written to try to fill that gap. In order to write it, we have literally traveled the world – from Silicon Valley to Siberia, from the Côte d'Azur of France to the middle of Korea, from South Australia to Andalucia. We have intensively studied more than a dozen technopoles worldwide: some long-developed and mature, some hardly even begun; some wildly successful, some apparent failures. In this book we try to tell how each began, how it has taken shape, what it is trying to achieve, and how well it has succeeded. And then we try to sum up our experience.

To start, however, we need to understand the true significance of the technopole phenomenon. These developments have not suddenly sprouted by some kind of accident, or just because of a passing fashion. On the contrary: they are deliberate attempts, by farseeing public and private actors, to help control and guide some exceedingly fundamental transformations that have recently begun to affect society, economy, and territory, and are beginning to redefine the conditions and processes of local and regional development.

THREE CONTEMPORARY ECONOMIC REVOLUTIONS

Technopoles in fact explicitly commemorate the reality that cities and regions are being profoundly modified in their structure, and conditioned in their growth dynamics, by the interplay of three major, interrelated, historical processes:

1 A technological revolution, mainly based in information technologies (including genetic engineering), at least as momentous historically as the two industrial revolutions based on the discovery of new sources of energy.[2]
2 The formation of a global economy, that is, the structuring of all economic processes on a planetary scale, even if national boundaries and national governments remain essential elements and key actors in the strategies played out in international competition. By a global economy we understand one that works in real time as a unit in a worldwide space, be it for capital, management, labor, technology, information or markets.[3] Even firms that are anchored in and aimed at domestic markets depend on the dynamics and logic of the world economy through the intermediation of their customers, suppliers, and competitors. The acceleration of the process of European integration and the creation of the new European Economic Area emphasize these tendencies towards globalization and interdependence in the world economy.
3 The emergence of a new form of economic production and management, that – in common with a number of economists and sociologists – we term informational.[4] It is characterized by the fact that productivity and competitiveness are increasingly based on the generation of new knowledge and on the access to, and processing of, appropriate information. As the pioneer work of Robert Solow and the subsequent stream of econometric research by the "aggregate production function" school of thought have shown, the last half-century has been characterized by a new equation in the generation of productivity, and thus of economic growth.[5] Instead of the quantitative addition of capital, labor, and raw material, typical of the function of productivity growth in both agrarian and industrial economies, the new economy, emerging in advanced industrial countries since the 1950s, has increasingly depended for its productivity growth on what econometric equations label the "statistical residual," which most experts translate in terms of the inputs from science, technology, and the management of information in the production process. It is this recombination of factors, rather than the addition of factors, that appears to be critical for the generation of wealth in our economy.[6]

Furthermore, the informational economy seems to be characterized by new organizational forms. Horizontal networks substitute for vertical bureaucracies as the most productive form of organization and management. Flexible specialization replaces standardized mass production as the new industrial form best able to adapt to the variable geometry of changing world demand and versatile cultural values.[7] This is not equivalent, as it is sometimes argued, to saying that small and medium businesses are the most productive forms of the new economy. Major multinational corporations continue to be the strategic nerve centers of the economy, and some of its most innovative actors. What is changing is the organizational form, both for large corporations and small businesses. Networks are the critical form for the flexible process of production. They include networks between large firms, between small and large firms, between small firms themselves, and within large firms that are decentralizing their internal structure, breaking it up in quasi-independent units, as so dramatically illustrated by the recent restructuring of IBM to offset its losses.

INFORMATION AND INNOVATION

These three processes are interlinked. The informational economy is a global economy because the productive capacity generated by the new productive forces needs constant expansion in a world market dominated by a new hierarchy of economic power, which is decisively dependent on information and technology, and is less and less conditioned – non-renewable energy sources temporarily excepted – by the cost of labor and raw materials.[8] The technological revolution provides the necessary infrastructure for the process of formation of the global, informational economy, and it is fostered by the functional demands generated by this economy. New information technologies are critical for the processes and forms of the new economy, on at least three levels.

First, they provide the material basis for the integration of economic processes worldwide, while keeping the necessary organizational flexibility for such processes. For instance, microelectronics-based manufacturing allows advanced standardization of parts of a given industrial product, produced in various locations, for assembly near the final market, while providing the necessary flexibility for producing in short runs or in large quantities, depending upon changing demand. The increasing integration of telecommunications and computing forms the technological infrastructure of the new global economy, as railways provided the material basis for the formation of national markets in nineteenth-century industrialization.

Second, industrial producers of new technologies have been the fastest-growing sectors in the world economy in the last 25 years, and in spite of business cycles they have certainly not yet reached their mature stage, driven as they are by constant innovation. Thus, such sectors play the role of growth engines for the development of countries and regions, with their greatest potential at the upper levels of the technological ladder, in industries characterized by a strict spatial division of labor between the innovation function, advanced fabrication, assembly, testing, and customized production. The more countries and regions are able to generate the development of these new, technologically-advanced industries, the greater their economic potential in the global competition.

Third, this information-driven technological transformation of the global economy requires a rapid modernization process of all sectors of the economy so that they are able to compete in an open economy. Thus, whether we consider the fate of nations, regions, or cities, technological diffusion becomes even more critical than the development of high-technology production.

In theory, such technological modernization could proceed quite independently from the design and production of advanced technological devices. However, empirical evidence shows that the technological potential of countries and regions is directly related to their ability to produce, indeed manufacture, the most advanced technological products, which constitute technological inputs that condition the improvement of products and processes in downstream industries.[9] This is for three reasons:

1 Technological innovation, and the application of such innovations, depends on the process of learning by doing, rather than off-the-shelf operation manuals. Thus, the greater the capacity of a country and region to design and produce advanced technological inputs, the greater its ability to adapt these technologies to productive processes elsewhere, creating a synergistic interaction between design, production, and utilization.[10]
2 All technical division of labor becomes, over time, a social division of labor. This means that in a world governed by competition through comparative advantage, countries or regions specializing in the production of inputs that are required by other industrial structures have a definite advantage. The technological component of products thus becomes a decisive dividing line in the trade between countries.[11] Following Ricardo's classical rule, Portugal specialized in wine and England in manufactured textiles; but, before long, the best port wine was the favorite drink of English gentlemen enriched by the product of their textile mills. In a similar manner, if Amstrad becomes a packaging and marketing device for Korean computer parts and ICL becomes a front store for Fujitsu, it may not be long before City of London stockbrokers will find that the trading language of their shopfloor has changed from English to Japanese, or perhaps Korean. The point is that the balance of competitive advantage in both these cases, and indeed in all cases in the long run, is with the technologically-advanced partner. Nations ignore this rule at their peril: importing other people's technologies in order to develop value-added services may employ a few people, but the resulting employment growth is likely to be both modest and highly volatile, as a comparison of the recent course of the Japanese and German economies with the British and American ones will amply demonstrate.[12]

In sum, the technological basis of countries and regions becomes critical for growth because, ultimately, the deficit in the balance of trade between high-value, high-technology producers, and low-technology, low-value producers creates an untenable disequilibrium. Research by Dosi and Soete on comparative trade patterns has demonstrated the fundamental role of high-technology manufacturing and the technological level of industrial sectors in international competitiveness.[13] Thus, if countries or regions do not generate sufficient surplus to import and adapt new technologies, they will be unable to afford the imports necessary for the modernization of their traditional industries.

3 Third, the culture of the technologically-advanced, information-based society cannot be productively consumed if there is no significant level of innovation in the social fabric. To be sure, the middle classes in developing countries can buy VCRs and personal computers. But only in a country, in a region, in a locality, where innovative informational processes are taking place can the generation of new ideas and new forms of organization and management occur in a creative manner. In other words, what is characteristic of the new informational economy is its flexibility, its productive adaptation to the conditions and demand of each society, of each culture, of each organization. To replicate the industrial organization of standardized mass production in the informational age, merely by buying the use of the technology without truly using its potential, is like using word-processing capacity to standardize the work of typing pools, instead of automating the process and upgrading secretarial work to the programming of more complex tasks.[14]

Technological innovation, the production of technologically-advanced devices, and technological diffusion cannot be entirely disjointed processes. Obviously not many regions in the world can excel in all three dimensions, and some inter-regional and international division of labor will always take place. However, no country or region can prosper without some level of linkage to sources of innovation and production. If this sounds an impossible task, it is because of an exceedingly simplistic notion about high-technology innovation and production. High technology is shorthand for a whole array of new products and processes that goes far beyond microelectronics, even if microelectronics was the original core of the technological revolution. Informatics (hard and soft), telecommunications, genetic engineering, advanced materials, renewable energy, specialty chemicals, information processing, bioelectronics, and so many other fields and subfields of technological innovation, advanced manufacturing, and technological services, offer so many opportunities that the scope of the new industrial geography, with its different levels of specialization and its diversity of markets, is broader by far than that generally accepted.

CITIES AND REGIONS: THE NEW ECONOMIC ACTORS

What we are witnessing, then, is the emergence of a new industrial space, defined both by the location of the new industrial sectors and by the use of new technologies by all sectors. At the same time, such new industrial space is globally interdependent, both in inputs and markets, triggering a restructuring process of gigantic dimensions that is felt by cities and regions around the world.[15]

Indeed, the most fascinating paradox is the fact that in a world economy whose productive infrastructure is made up of information flows, cities and regions are increasingly becoming critical agents of economic development: as Goodman puts it, the last entrepreneurs.[16] Precisely because the economy is global, national governments suffer from failing powers to act upon the functional processes that shape their economies and societies. But regions and cities are more flexible in adapting to the changing conditions of markets, technology, and culture. True, they have less power than national governments, but they have a greater response capacity to generate targeted development projects, negotiate with multinational firms, foster the growth of small and medium endogenous firms, and create conditions that will attract the new sources of wealth, power, and prestige. In this process of generating new growth, they compete with each other; but, more often than not, such competition becomes a source of innovation, of efficiency, of a collective effort to create a better place to live and a more effective place to do business.

In their search for the new sources of economic growth and social well-being, cities and regions are stimulated both negatively and positively by comparative international experience. Those areas that remain rooted in declining activities – be they manufacturing, agriculture, or services of the old, non-competitive kind – become industrial ruins, inhabited by disconsolate, unemployed workers, and ridden by social discontents and environmental hazards. New countries and regions emerge as successful locales of the new wave of innovation and investment, sometimes emerging from deep agricultural torpor, sometimes in idyllic corners of the world that acquire sudden dynamism. Thus, Silicon Valley and Orange County in California; Arizona, Texas, Colorado, in the western United States; Bavaria in Germany; the French Midi, from Sophia-Antipolis via Montpellier to Toulouse; Silicon Glen in Scotland; the electronics agglomeration in Ireland; the new developments in southern Europe, from Bari to Malaga and Seville; and, above all, the newly industrializing countries of Asia (South Korea, Taiwan, Hong Kong, Singapore, Malaysia) that in two decades have leapt straight from traditional agricultural societies – albeit with high levels of literacy and education – to being highly competitive economies based on strong electronics sectors.

Such role models, both positive and negative, have a dramatic influence on the collective consciousness of countries, regions, and localities, as well as on the development projects of their respective governments. Many regions in the industrialized and industrializing world have dreamed of becoming the next Silicon Valleys, and some of them went headlong into the business. A hasty, hurried study by an opportunistic consultant was at hand to provide the magic formula: a small dose of venture capital, a university (invariably termed a "Technology Institute"), fiscal and institutional incentives to attract high-technology firms, and a degree of support for small business. All this, wrapped within the covers of a glossy brochure, and illustrated by a sylvan landscape with a futuristic name, would create the right conditions to out-perform the neighbors, to become the locus of the new major global industrial center. The world is now littered with the ruins of all too many such dreams that have failed, or have yielded meager results at far too high a cost. Indeed it would almost seem, as some scholars have argued, that the whole world has become gripped by "high technology fantasies," which signify virtually nothing.[17]

And yet, the late twentieth century has seen an undeniable major redistribution of technological innovation and industrial entrepreneurialism; the world's economic geography has very fundamentally changed.[18] The fact that local and regional governments have stampeded to join the new model of development does demonstrate their perception that we are indeed in the midst of a transition to a new productive form, and that managing the process requires institutional initiatives working through and upon the market. True, image-making and high-technology ideology are pervasive elements of this new brand of regional policy. But this is because, for good or for ill, image-making has become a central basis for successful competition in our latter-day economy and culture. By trying positively to harness the new technologies to their own ends, localities and regions are asserting their control over events; they are vigorously denying that they are condemned to live within the old logic of spatial divisions of labor, that locks them into particular functions determined by events long ago. As in any process of competition and entrepreneurial initiative, there will be creative destruction – as well as (with due apologies to Schumpeter) destructive creation, when a region bases its future on military-related high-technology production.[19] Yet this drive to innovate and invest is successfully building new industrial spaces, and is thus producing a new and quite extraordinary wave of worldwide reindustrialization that denies the myths of postindustrialism.

TECHNOPOLES AND MILIEUX OF INNOVATION

This effort to innovate and develop *de novo* very often takes the form of creating and nurturing what we have called "technopoles." We need now a more precise definition; under this name we include various deliberate attempts to plan and promote, within one concentrated area, technologically innovative, industrial-related production: technology parks, science cities, technopolises and the like. Our study will try to assess how these different developments perform (or fail to perform) their role as engines of the new round of economic development and as organizing nodes of the new industrial space. Comprehension of the technopole phenomenon has been so blurred by political, ideological, and business biases, that any serious study must start from a careful empirical analysis of how these centers are created and developed, and of the factors that account for their differential success, according to a set of criteria that must be established at the start.

However, our study of the technopole phenomenon soon led us to the conclusion that we should not be constrained by the boundaries artificially set by the promoters of the technopole idea. In other words, we must go back to seek the historic sources of inspiration of the technopole strategy. We should look into the formation and operation of those innovative industrial complexes that first changed the dynamics of world competition: Silicon Valley, Boston's Route 128, the Los Angeles military-industrial complex, the Tokyo industrial district, the post-1945 industrialization of Bavaria.

Furthermore, in our inquiry we came to the conclusion that most of the older major metropolises, such as Paris or London, indeed remain among the major innovation and high-technology centers of the world, and certainly of their respective countries. Thus, the focus of our study gradually shifted from exclusive concentration on the deliberately-planned technopoles, to extend both to their implicit role models and their true parents, that is, to the semi-spontaneous technopoles and to the giant metropolitan technopoles that cast such a huge shadow over the new start-ups and their aspiring imitators. And that in turn led us to consider, in all its complexity, the meaning of that pregnant but elusive concept, milieux of innovation.

By milieux of innovation we understand the social, institutional, organizational, economic, and territorial structures that create the conditions for the continuous generation of synergy and its investment in a process of production that results from this very synergistic capacity, both for the units of production that are part of the milieu and for the milieu as a whole.[20]

Developing such milieux of innovation has now become a critical issue for economic development, and a matter of political and social prestige. And thus image-making, industrial projects, state policies, and the new economic geography mix in a confusing game of simultaneous doing and labeling. Any attempt to analyze the interaction between technological development, industrialization, and regional development on the basis of international experience must start with a clear distinction between the various kinds of realities to which terms like technopoles or technology parks, or any of the other labels, refer. This is not just a semantic question, since each type of technopole must be analyzed and evaluated according to the implicit or explicit goals it is trying to achieve.

A TYPOLOGY OF TECHNOPOLES: OUTLINE OF THE BOOK

While the eventual purpose of this book is to analyze and thus understand the process of formation of innovative industrial milieux, leading to truly dynamic regional or local economic growth, we start with a more modest aim: to tell the story as it actually is, in terms of the case studies of real places, before we proceed into the analysis. So the typology of technopoles we propose here is an empirical one: these are the various kinds of technologically innovative milieux that we actually find in today's world. Some may have more ambitious goals, some more modest ones, but they are all presented here as they are: as specific forms of territorial concentration of technological innovation with a potential to generate scientific synergy and economic productivity. Our taxonomy, accordingly, arises from the facts of the international experience in all its varied forms.

The first type of technopole consists of industrial complexes of high-technology firms that are built on the basis of innovative milieux. These complexes, linking R&D and manufacturing, are the true command centers of the new industrial space. Some are created entirely out of the most recent wave of global industrialization, characterized by the new high-technology firms; the most prominent worldwide is Silicon Valley, which we have accordingly selected so that we may study and report on the actual history of an industrial myth. Other new complexes, however, develop out of old industrial regions which go through a process of transformation and reindustrialization; the most important example is Boston's Route 128, which accordingly becomes the second of our case studies.

These new techno-industrial complexes arise without deliberate planning, though even there governments and universities did play a crucial role in their development. But other experiences are indeed the result of conscious institutional efforts to replicate the success of such examples of spontaneous growth. Thus, most of our analytic efforts will focus on the experiences of planned techno-industrial development.

The next type of technopole we distinguish, accordingly, can broadly be called science cities. These are strictly scientific research complexes, with no direct territorial linkage to manufacturing. They are intended to reach a higher level of scientific excellence through the synergy they are supposed to generate in their secluded scientific milieux. We have chosen to study four major cases of deliberate attempts to create scientific excellence by concentrating human and material resources in such a secluded science center, in four very different contexts: the Siberian city of Akademgorodok, the major Japanese experiment at Tsukuba, the Korean creation at Taedok, and the new concept of the multinuclear science city currently under development in the Kansai region of Japan.

A third type of technopole aims to induce new industrial growth, in terms of jobs and production, by attracting high-technology manufacturing firms to a privileged space. Innovation functions are not excluded from such projects, but they are mainly defined in terms of economic development. We label them "technology parks," because that is the most usual way in which they describe themselves. However, the name should not obscure the reality, which is a deliberately-established high-technology business area, resulting from government- or university-related initiatives. In this quite loose category, we have analyzed three experiences, ranging from the most tightly-planned government-originated park, through a mixed scheme, to the loosest university initiative: the cases of Hsinchu in Taiwan, Sophia-Antipolis in France, and Cambridge in England.

Fourth, we have analyzed the design of entire technopolis programs as instruments of regional development and industrial decentralization. Here the choice was obvious, since there is only one major such program worldwide: the Technopolis program in Japan.

Having looked at these planned cases, and having reflected on them, we came to a conclusion that surprised us. Despite all this activity, it remains true that over the years and decades most of the world's actual high-technology production and innovation still comes from areas that are not usually heralded as innovative milieux, and indeed may have few of their physical features: the great metropolitan areas of the industrialized world. So we determined that we should study them as quintessential innovative milieux.[21] We found a fundamental distinction between the old metropolises that kept their leading technological role (Tokyo, Paris, London), the metropolises that lost their role as advanced manufacturing centers (New York and Berlin), and the newly-arriving technological-industrial metropolises that in fact took their place (Los Angeles and Munich).

Finally, we decided that it might be interesting to make a progress report on two current attempts to create innovative milieux, in which the authors are themselves involved: Adelaide's Multifunction Polis and Seville's Cartuja '93, two antipodean experiences that should yield valuable lessons for any other place, or other group of people, interested to join the great adventure of planning the territories of the next technological age.

These works in progress conclude our case studies. In two final chapters, we try to draw the threads together. Chapter 9 seeks to distill the lessons that these very disparate experiences appear to teach us. And Chapter 10 builds on it to make some tentative suggestions for a policy of technopole-building. In these chapters, necessarily, we are somewhat speculative and judgmental. We are dealing with some of the largest and certainly some of the potentially most important projects anywhere in the world today. These are our judgments on them, and not everyone may necessarily agree with them. We hope that at least they will trigger a debate, and perhaps further studies. If they do, then our purpose will have been satisfied.

2

SILICON VALLEY: WHERE IT ALL BEGAN

Silicon Valley has a guaranteed place in history as the original industrial core of the revolution in information technologies. While its reputation is based on the basic fact of a concentration (in 1989) of some 330,000 high-technology workers, including 6,000 Ph.D.s in engineering and science,[1] it stems also from the saga of Silicon Valley, hailed worldwide as an heroic model of innovation in the service of dynamic economic growth.

This 40-mile by 10-mile (70-kilometer by 15-kilometer) strip in the peninsula south of San Francisco, stretching from Palo Alto to the southern suburbs of San Jose (Figure 2.1), has become the popular epitome of entrepreneurial culture, the place where new ideas born in a garage can make teenagers into millionaires, while changing the ways we think, we live, and we work. It is also seen as living proof of the fundamental relationship between science and economic development, a process that emphasizes the role of universities and research as driving forces of human progress. Last but not least, Silicon Valley embodies the new power emerging from new technologies: the battles of future wars are fought in its electronics laboratories; competitiveness in the world economy largely depends on access to the kind of technological excellence that is so richly concentrated here; industrial and military spies make a living out of it; companies line up to establish joint ventures with its innovative firms; presidents, ministers, and dignitaries come in pilgrimage here, in well-publicized delegations that aim to capitalize the visit in social prestige or political votes back home.

In Santa Clara County, the area's true geographical name, and in its immediate surroundings were concentrated in the 1980s about 3,000 manufacturing electronics firms, 85 percent of which had fewer than 50 workers (Table 2.1). Another 3,000 firms in the area provided necessary producer services, and 2,000 other firms were engaged in high-technology activities, giving a total of 8,000 firms in the complex.[2] Yet, as late as 1950, Santa Clara County was mainly an agricultural area, with only 800 manufacturing workers, most of them in food processing plants.[3] During the 1970s one new firm was created every two weeks, and 75 percent of them survived for at least six years, a much higher rate of resilience than American companies

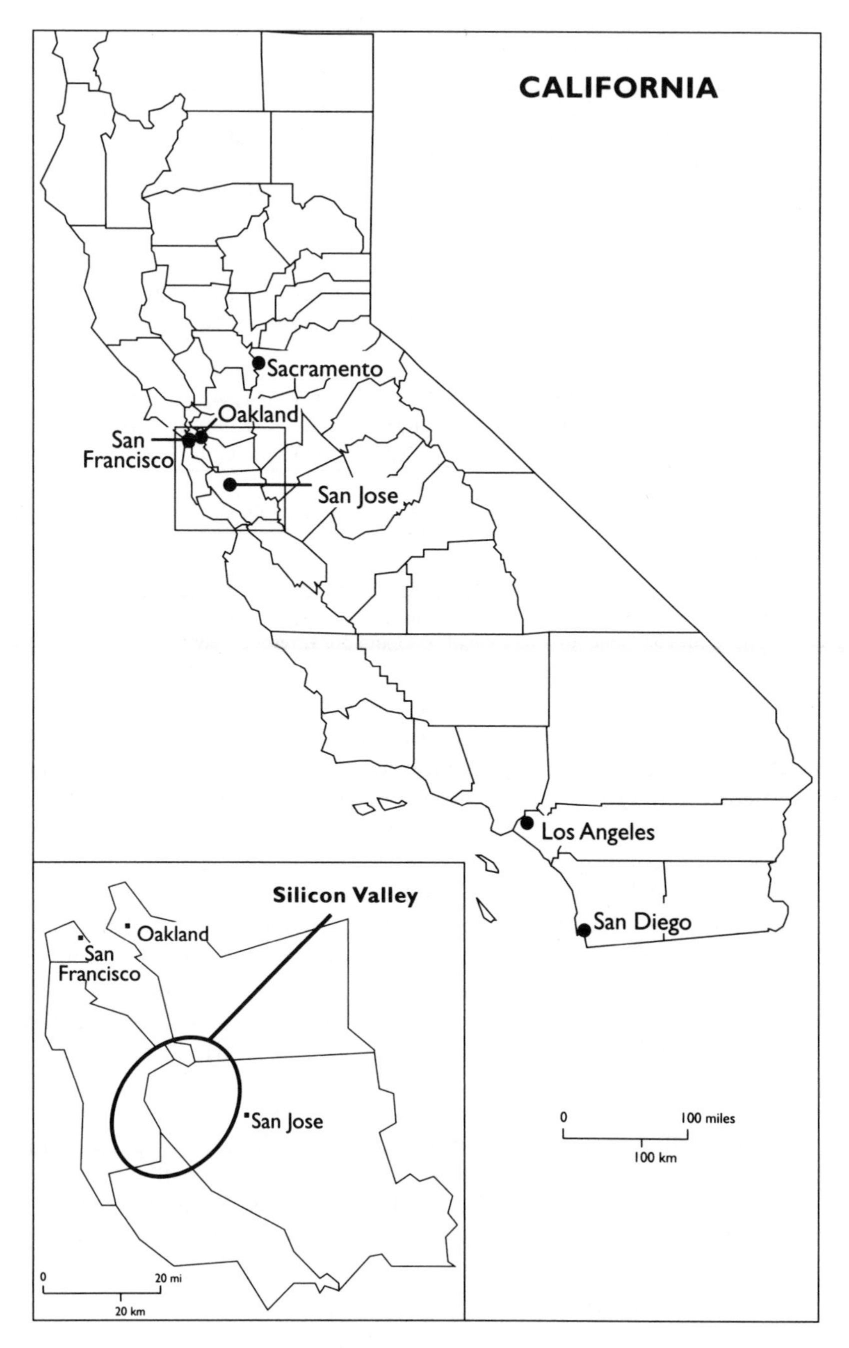

Figure 2.1 Silicon Valley: general location
Source: Rand McNally World Atlas, 1992

Table 2.1 Silicon Valley: employment structure

Employment	*SIC*	*1959*	*1965*	*1970*	*1975*	*1980*	*1985*
Computers	3,573	0	0	8,938	19,902	52,738	56,126
Other office machines	357*	0	0	979	1,869	2,582	2,748
Communications	366	895	5,027	7,271	10,043	19,603	29,677
Semiconductors	3,674	0	4,164	12,290	18,786	34,453	47,069
Other electronic components	367*	4,295	4,619	14,174	11,622	25,472	23,731
Missiles/parts	372	0	0	2,274	0	0	750
Instruments	38	328	1,202	2,567	14,646	24,912	19,382
Drugs	283	0	282	0	750	1,976	1,954
Software/data processing	737	0	0	0	3,887	7,813	15,368
IC labs	7,391	118	2,193	1,978	1,642	3,856	6,133
Electronic wholesale	5,065	131	693	1,107	2,092	3,703	9,179
Computer wholesale	5,086	199	243	373	620	2,005	2,807
Total high-tech employment		5,966	18,423	51,951	85,859	179,113	214,924
Total manufacturing employment		61,305	88,038	131,613	154,126	256,437	272,332

Source: US Bureau of the Census, County Business Patterns, selected years
Note: * SIC Codes 357 and 367 exclusive of 3573 and 3674, respectively

overall.[4] During the 1980s, in spite of a severe downturn in the computer industry in 1984–6, the dynamism of the regional economy continued unabated, with a growth of 35 percent in total employment in 1980–9. Since 1970, median family income in the area has been the highest in California and one of the highest in the United States.[5]

This vibrant economy is based on an extraordinary capacity for innovation: most of the key inventions in microelectronics and computing have originated in Silicon Valley, including the co-invention of the integrated circuit, the planar process, the microprocessor, the Unix system, and the development of the personal computer.[6] The fact that such new industrial power could emerge from an area without any previous manufacturing basis or business tradition – San Francisco was always clearly external to the Silicon Valley complex – has struck the imagination of policy makers and the media the world over. However, it is precisely because the story is extraordinary that it must be carefully and analytically reconstructed. Only thus is it possible to avoid the ideological pitfalls that, inevitably, are linked to the legend. For, in reality, Silicon Valley cannot be reduced simply to a bright illuminated billboard proclaiming the virtues of free-market ideology.

THE SILICON VALLEY STORY

While the Silicon Valley story has been told many times – although, oddly enough, rarely in scholarly research, with some notable exceptions[7] – it is still useful to recount the facts in historical sequence, in order to be able to comprehend both the area's uniqueness and its potential for generalization. It is helpful to summarize the development of Silicon Valley as a major industrial technological center in a sequence of stages:

1 The historical precedents of technological innovation in Silicon Valley, from the early twentieth century.
2 The creation of its high-technology industrial basis in the 1950s around the Stanford Industrial Park.
3 The growth of innovative microelectronics firms in the 1960s, on the basis of spin-offs of the first generation of firms, and with the support of the Department of Defense's electronics-based programs.
4 The consolidation of semiconductor merchant producers in the microprocessor era, together with the launching of the personal computer era during the 1970s.
5 The growing domination of the computer industry, the internationalization of the industrial structure of the Valley, and a new round of innovative spin-offs in the 1980s.

This historical sequence is crucial to understanding Silicon Valley. For, though the factors that are generally associated with its growth and success have all been important, individual elements have been important at different periods and with different intensities in each period – a fact that makes all the difference in the world for the purpose of analysis and generalization.

First of all, in spite of its lack of a previous industrial basis, the area did have a significant research tradition in electronics going back to the invention of the vacuum tube by De Forest in 1912, in a firm, the Federal Telegraph Company, that was formed by a Stanford University graduate with support from the University.[8] During the 1920s Stanford maintained a tradition of excellence in electrical engineering around Dr Harris Ryan, many of whose students stayed to work in electronics in the area, creating their own firms.

But the crucial link between the early stage of electronics and the formation of Silicon Valley was provided by Frederick Terman – the man who, more than anyone else, can be considered the central figure of the story. Frederick Terman was the son of a Stanford psychology professor and grew up on the campus. He studied chemistry and electrical engineering at Stanford, then enrolled in the doctoral program in electrical engineering at the Massachusetts Institute of Technology (MIT), the best in the country at that time. Completing his degree, he was appointed to the MIT Faculty, but he contracted tuberculosis, and decided to stay in Palo Alto because the climate was better for his health, becoming Professor of Radio Engineering at Stanford. Thus, given the centrality of his role to the development of Silicon Valley, we may say that climate did play an important place in its emergence as a high-tech center – but not so much for its contribution to the quality of life, as we will see, as for its contribution to the health of Terman. He was promoted head of the communications laboratory at Stanford, and later Dean of Electrical Engineering, before becoming Provost and Vice-President of the whole University during the 1950s.

Convinced as he was of the critical need to link University and industry, he used all his connections, his influence, and sometimes his own money to encourage his best graduates to start up electronics firms, a high-risk venture in the 1920s and 1930s.[9] Among his students were Charles Litton, who in 1928 founded Litton Engineering Laboratories, and two particularly gifted graduates, William Hewlett and David Packard. Terman protected them, helped them with their research, persuaded them to set up a commercial firm to exploit their research, and lent them $1,538 to start the company in 1938; by 1942, company sales reached $1 billion. World War Two, with its appetite for electronic devices, gave a decisive impulse to Hewlett-Packard and to the other start-ups around Stanford, while Terman himself spent his time at MIT managing a high-technology military project.

Upon his return to Stanford he went on to build up a major program in electrical engineering, modeled after the much superior East Coast programs, with support from local corporations. But his main goal was still to be able to diffuse R&D results in the industrial world. Since Stanford was a land-rich university, from its original grant, in 1951 Terman decided to use this asset to create the Stanford Industrial Park, the true ancestor of all the world's future technopoles. Terman leased the land to firms with very advantageous terms, on the basis of their excellence in electronics technology and of their close contacts with the University.

The first company to move to the Park was Varian, another Stanford spin-off which had Terman on its Board of Directors. Hewlett-Packard moved to the Park in 1954. By 1955 there were seven firms, by 1960, 32, by 1970, 70, and by the 1980s, 90, with about 25,000 workers. By 1954 Terman's dream of a "community of technical scholars," based upon the relationship between the University and private industry,[10] had become a reality, constituting the nucleus of what would be Silicon Valley. However, for all the efforts in upgrading the West Coast universities, they were clearly behind the traditional East Coast centers in electronics research until the 1960s. Thus, technology transfer from these centers became a necessary condition for Silicon Valley to develop a base in cutting-edge innovation.

Here enters the story the other godfather of the legend, Nobel Prize winner William Shockley, co-inventor of the transistor in 1947 with a team at Bell Laboratories in New Jersey. Shockley left Bell in 1954 to form his own company to commercialize his research.[11] He first tried the Boston area but because of the lack of support of the large firms there (Raytheon rejected his proposal), he moved to Palo Alto – because, among other reasons, his aged mother lived there. In 1955 he founded Shockley Semiconductors

Laboratory in Mountain View, near Palo Alto. Young electronics graduates responded immediately to his first job announcement, and he recruited the eight brightest; they would all become multimillionaires and technological innovators, including Robert Noyce, the co-inventor of the integrated circuit in 1957.

However, Shockley was as brilliant a scientist as he was a bad businessman and generally unpleasant person (he was a self-proclaimed racist), regarded as difficult to work with. His stubbornness led to his commercial demise, since he insisted on working on four-layer diodes and refused to move into silicon transistors, as his young disciples advised him to do. Thus, in 1957, the eight young engineers left Shockley and founded a new company, Fairchild Semiconductors, then the only transistor firm to work exclusively in silicon.

Within a few years, Fairchild was credited, mainly through the work of Bob Noyce, as the co-inventor of the integrated circuit (IC), and the inventor of the planar process, the critical manufacturing technology required actually to produce ICs. Fairchild attracted the best young talent in microelectronics but, at the same time, could not retain it because of their ambition and dynamism. By 1965, 10 new firms had been created by former Fairchild engineers.[12] Furthermore, when Fairchild was taken over by an East Coast parent corporation unable to understand the local innovative dynamics, its founders also left the firm, leaving it an empty shell.

Thus, about one-half of the 85 major American semiconductor firms were created as direct or indirect spin-offs from Fairchild, which thus became the major mechanism for the diffusion of technological know-how throughout the area.[13] Among Fairchild's spin-offs are Intel (created by Bob Noyce in 1968), National Semiconductors, Signetics, Amelco and Advanced Micro Devices, all leaders in the industry today. All these companies set up shop in Silicon Valley, moving to the next available nearby location to continue work on the same programs and almost with the same people. Of the 45 American semiconductor firms created between 1959 and 1976, 40 were located in Silicon Valley.[14]

However, for this process of spin-off to be successful, other factors intervened in the 1960s; and these made the Silicon Valley story more complex. The first was the dramatic expansion of military demand for electronic devices in the late 1950s and the 1960s, as the aerospace program took shape in the wake of the shock provoked in the American establishment by the launching of the first Sputnik.[15] For instance, in 1959 Fairchild was awarded a $1.5 million contract to provide the transistors for the Minuteman missile, and in 1963 the integrated circuits for the Apollo spacecraft's guidance computer. In the late 1950s the share of military markets in total shipment of semiconductors reached the 70 percent level, and it oscillated around 50 percent during the 1960s, with the defense market being concentrated in the higher layers of the technology.[16]

Both the Defense Department and NASA paid high prices for the most innovative technologies, which were also the riskiest in terms of investment; thus, they played the role of subsidizers of R&D for Silicon Valley firms. In addition, the practice of second-sourcing and technology sharing by military departments, to ensure the timely provision of the required devices, led to rapid technology diffusion among firms, and to lack of proprietary control over the inventions. Indeed, unless they were classified as military secrets, the Defense Department required the public diffusion of the discoveries realized through its funding.[17] Thus, military and aerospace demand provided the first mass market for the infant microelectronics industry concentrated in Silicon Valley, and facilitated the necessary capital for high-risk investment, while making possible the diffusion of technological breakthroughs among the firms.

But Silicon Valley firms were able to take advantage of the military bonanza because of their extraordinary versatility, entrepreneurialism, and high level of mutual interaction. Scholarly research, in particular the pioneer work by AnnaLee Saxenian, has emphasized the decisive role of social networks and of a shared culture of innovation in the formation, development, and continuing vitality of Silicon Valley.[18] The Fairchild spin-offs were often projected, discussed, and decided in a nearby restaurant in Mountain View, Walker's Wagon Wheel Bar and Grill, frequented by the company's engineers.[19] The constant circulation of talent from one firm to another made it literally impossible to maintain proprietary rights over each innovation. The only way out of the problem was for each company to accelerate its own path of innovation, eventually giving way to new spin-offs, in an endless process of extraordinary technological and industrial self-stimulation.

But while scientists and engineers would change jobs easily, always looking for better opportunities and for more exciting research challenges, they generally stayed in Silicon Valley because they kept alive their informal networks, based on face-to-face interaction over common technical or professional issues. These informal networks, as Saxenian has demonstrated in her doctoral dissertation,[20] constituted the very basis of the process of innovation in Silicon Valley, and they increased in complexity and importance over time. They were simultaneously channels of communication of technological innovation, forms of organization of the job market, and the material basis for the formation of a culture that emphasized the values of technological excellence and free-market entrepreneurialism, while serving to transmit the role models of the Valley: the brilliant young engineers who achieved wealth and fame through technical expertise and social irreverence.

These networks supported the development of another key ingredient of Silicon Valley, the venture capital firms, that were decisive in providing finance for the development of electronic firms outside the original narrow ground of military markets. But, against the common view on the matter, it seems that venture capital firms did not originate in the San Francisco financial markets, but from the wealth generated in Silicon Valley itself.[21] In fact, there were engineers and businessmen among the first wave of electronics firms who invested their money in the next round of start-ups, having verified from their own experience the feasibility of the process, and feeling competent enough to judge the possibilities of the proposed new firms. Thus, although in the 1980s major financial institutions opened up shop in the venture capital market in the Valley, earlier – in the 1960s and early 1970s – the Valley's own social networks created a self-support system of finance, reinvesting part of their wealth in fostering the next generation of entrepreneurs.

By the mid-1970s Silicon Valley had developed its social networks, its industrial basis, its supporting financial and service activities, and its professional organizations, to the point of constituting an innovative milieu able to absorb and propel into the market key innovations that were not of its own. That was particularly the case of the product that changed the world and the Valley, opening up a new industrial era for the region: the personal computer.[22] The personal computer was first produced in 1974 – in Albuquerque, New Mexico, by an engineer, Ed Roberts, working out of his small calculator company, MITS – in the form of a model named "Altair." Altair, in spite of being a primitive machine, was an instant commercial success. But its main impact was to mobilize the informal network of computer hobbyists that was already in existence in the San Francisco Bay Area.

The core of this network was the Home Brew Computer Club, formed by young electronics engineers and computer lovers, meeting regularly to exchange information and discuss developments in the field. The network included people like Steve Wozniak, the future inventor of Apple; Bill Gates, the software guru, founder of Microsoft; and other young visionaries who would go on later to start 22 companies, including Apple, Microsoft, Comenco, and North Star. This was the network that took up Roberts's example of a personal computer, out of a romantic vision of giving computer power to the people, while established companies, including Silicon Valley companies such as Hewlett-Packard, rejected the first technical proposals.

The most widely publicized story of technological audacity and business imagination came out of this process: the development of the Apple Personal Computer by two school drop-outs in their early twenties, working out of their garage in Menlo Park in the summer of 1976. Steve Wozniak was the designer of the computer, and Steve Jobs the business genius who sensed the commercial potential of their product and went on to build a company that jumped from zero to $100 million sales in four years, to become a decade later a truly multinational corporation. It is important to notice that Jobs and Wozniak were only able to start the company because a former

Intel executive, Mike Markkula, came into the project as a third partner, lending them $91,000.[23]

It is this high-risk funding by individuals who were knowledgeable about the trade, and who shared and understood the culture of the innovators, that made possible the endless birth of new firms in Silicon Valley. When in 1981 IBM introduced its own PC, making official the importance of the new market, Silicon Valley networks started to generate computer companies, both in hardware and software, as well as computer services business, making computers by the mid-1980s the most important activity of the region, even surpassing semiconductors. In addition to Apple, and to the continuing expansion of Hewlett-Packard, a new computer company, Sun Microsystems, created in the early 1980s out of the same social networks, became another example of the capacity of Silicon Valley to generate major companies in a few years, out of new ideas and new technological breakthroughs, in the case of Sun mainly due to the contribution of another young computer guru, Bill Joy.

However, when in 1984–6 Japanese competition and a world downturn in the computer industry struck the region, forcing the lay-off of over 21,000 workers, many observers concluded that the aging industrial structure of Silicon Valley had peaked, and forecasted a slowing-down of the innovation drive. Yet, in the second half of the decade, the social networks in the area continued to act as magnets to information and capital from all over the world, generating new spin-offs, starting up new companies, and diversifying and making more complex the pre-existing industrial structure.[24] While the established semiconductor companies lost ground to Japanese competition in high-volume, standardized production, a new wave of companies, both in semiconductors and computers, went on to develop a new flexible production system, concentrating on high-value customized devices, and upgrading their technological level to edge off the competition.

Silicon Valley now became increasingly specialized in the high level of technological production in microelectronics and computers, with companies automating their manufacturing plants and/or moving them to other cheaper areas in the United States while keeping in the Valley the high-level functions of R&D, design, and advanced manufacturing. Changing networks of firms, specializing in various operations, and deepening their technological leadership at each stage of the production process, were the source of a new round of innovation and industrial growth that kept Silicon Valley ahead of the competition and growing, economically, demographically, and territorially, at the very time that other high-technology regions, such as Boston's Route 128, were hurt by the decline of military markets.

Thus, from the mid-1970s Silicon Valley became a self-sustaining innovative milieu of high-technology manufacturing and services, generating its own production factors: knowledge, capital, and labor. Universities, including Stanford, San Jose State, Santa Clara, and to some extent the relatively distant Berkeley, continue to be critical in providing the labor market with well-trained engineers and scientists. However, their role as sources of R&D has substantially declined in comparison to the endogenous research capacity of the industry, although cooperative research programs continue to link universities and firms.

But, for the innovative milieu to become self-sustaining, it had to be supported for a long period by the set of interactive elements whose development we have presented here. And for such elements to cluster together into a synergy-generating process, they have to be supported by a specific local culture both resulting from the high technology industry and contributing to its innovative capacity.

THE SILICON VALLEY CULTURE

Technological revolutions have always been associated in history with the emergence of specific cultures.[25] Such cultures are essential ingredients of the ability to innovate and to link innovation to the applications most valued in a given society, from the building of cathedrals to worldwide commercial sales or to the mastery of military power. The territorial concentration of innovation processes in certain core areas seems to be a prerequisite for the development of such culture and for the positive interaction between technological innovation and cultural change.[26]

In the case of Silicon Valley, there is indeed a strong cultural specificity in the values and lifestyle of executives, engineers, technicians, and skilled workers that forms the human basis of this leading milieu of innovation. Some of the features of this culture do not fit entirely with the virtues heralded by the legend, while others in fact do. There is little systematic scholarly research on the subject, so that our analysis is necessarily tentative. Yet, there is enough survey information, including journalistic reports and trade books written by scholars, to provide the basis for some broad characterization of the patterns of social values and behavior in the Valley.[27] We have used in particular a very important survey on a representative sample of Santa Clara County workers, conducted by the major local newspaper, the *San Jose Mercury News*, in August–September 1984,[28] a date prior to the crisis that struck the industry, thus modifying some of the classical social patterns in the area.

While accepting the risk of excessive schematism, we could synthesize the predominant Silicon Valley culture by nine interrelated features:

1 *The centrality of work*. Silicon Valley, like all major industrial centers, is as far as can be from the "laid-back" Californian image. For 49.2 percent of people surveyed "what they do at work is more important than the money they earn," and 38.7 percent said that their "main satisfaction in life comes from their work." Fifty-nine percent claimed to be "very satisfied" with their jobs, against 46.7 percent for the whole of the United

States. People do indeed work: 30 percent work between 41 and 50 hours a week; 10.4 percent spent more than 51 hours working; 28 percent take work home at least once a week. Job satisfaction increases with the number of hours worked.

Thus, hard, intense work is the basic feature of life for Silicon Valley producers, particularly for the most skilled segments of the population. As in the case of other technological-economic revolutions, the drive to produce and to successfully compete is the basic source of the new social organization. And, as in the previous industrial revolutions, work and the workplace tend to be the primary focus of social activity.

2 For the technical-professional component of the employed population (accounting roughly for 50 percent of high-technology workers) there is also a *positive feeling towards work as the opportunity for innovation*. They have the ideology of the innovators, expressed in the feeling of being in the cutting edge of technology and sensing the importance of it.

3 *Entrepreneurialism* is a fundamental feature of the culture, in spite of the fact that the majority of professionals and engineers work for large companies. Yet, the role models continue to be the young leaders of start-up companies who became millionaires out of their capacity to innovate and of their audacious attempts to create new firms. This culture provides the ground for endless spin-offs that have nurtured the durability of innovation of the Valley, in spite of the efforts of companies in the 1980s to slow down the very process that had been their origin, in a futile attempt to close the door once they were established in the market.

4 Another key cultural attitude is *aggressive competition*, both between individuals and between firms. There is an all-out struggle to keep ahead, leading to loose moral standards in professional relationships; this is something that seems to be characteristic of many historical contexts of major innovation, when the frontier spirit becomes cut-throat competition. Thus, 36 percent of workers surveyed think that "their co-workers" falsify their career histories. Forty-two percent think that "their co-workers" take material or computer time from the company without authorization; 16 percent believe that use of company secrets for personal gains occurs frequently; and 55 percent of high-tech workers think that "some people will do anything to get ahead." It is precisely this personal drive, invested in revolutionary technologies applied to a strategic industry that leads to the acceleration of the innovation process in the area, either through the existing firms or through their spin-offs.

5 As one could imagine, this is a culture of *extreme individualism*, possibly stimulated by the continuing immigration to the region of thousands of young professionals from all over the world, attracted by the "Silicon Rush." In 1984, 31 percent of the workers surveyed had never married; 15 percent were divorced; and only 20 percent of the labor force was over 45 years of age. Such an individualistic pattern has direct consequences

on housing markets, on the school system, on traffic behavior, on leisure, and on politics, governments of all kind being universally distrusted and taxes being considered a crude assault on the individual citizen. The "free rider" ideology blossoms in Silicon Valley on the basis of a highly educated, often single, mainly out-of-state immigrant population that strives for the high rewards in which it believes. The rewards are actually there – if not for everybody, at least for enough people to make everybody believe in the possibility of reaching them.

6 Although not a cultural element in itself, it is important to emphasize that such cultural expressions rely on a material basis: *the affluence of the area*. To be sure, there are also dark sides to this prosperity: poverty, discrimination and exploitation are present in Silicon Valley as in all class societies. Yet, on the average, and in relative terms to other areas, including high-technology areas, there is an undeniably high standard of living for the majority of the population. Such reality leads to high expectations from most people, who feel it is possible to make a good living while still young. A direct consequence of such a feature is the inability to pay attention to, or even to understand, those left out of the affluent group, reinforcing individualism, and digging even deeper the trenches of urban segregation.

7 Merciless individualistic competition, and the relentless drive for work and innovation have a major cost: *technostress*, as it is called in Silicon Valley, meaning social and psychological stress in all their manifestations. Thirty-eight percent of all surveyed workers, 42 percent of women, and 43 percent of professionals, said that job-related stress affected their lives off the job. Job-related stress is associated in the opinion of most experts with widespread social ills in the area:[29] alcohol and drugs (one-third of those surveyed think "their co-workers" often use drugs at work); family disruption (with one of the highest divorce rates in the United States); frequent emotional problems for children, etc. Furthermore, in addition to stress in itself, there is also a culture of stress, that is, a value system in which extreme stress has become part of the lifestyle as the necessary price to pay to be on top of the world. Mechanisms to deal with stress were generated by the Valley's firms and institutions, becoming an integral part of the local culture. The following two are the main such mechanisms.

8 The emergence of *corporate subcultures*, of which Hewlett-Packard is the most distinctive. Feelings of company loyalty are very strong for 65 percent of the surveyed workers (against 46 percent at the national level). Firms tend to stimulate such feelings of membership via recreational activities, flexible working schedules, and informal styles of personal interaction. The aim is to offset high labor turnover and to retain in the company the brain power that represents the main asset of a research-based firm. Thus, the pattern of cut-throat competition we have described

is somewhat smoothed by team work and by interpersonal cooperation in the workplace, a device providing psychological support among equally workaholic individuals.

9 Another major mechanism to relieve stress is what could be labeled *compensatory consumption*, at least for the affluent half of the population. Because of the drive for innovation and the search for immediate rewards, helping to release stress, consumption styles tend to emphasize extravagance, experimentation, and lavish, *nouveau riche* behavior. The "hard work–hard play" syndrome of Silicon Valley is a way of materializing in the short term the rewards workers expect from the effort they put into innovation and competition. The classic deferred gratification pattern of the Protestant ethic is replaced by an immediate gratification pattern, coupled with the importance of work as a goal in itself. So, consumption is not in this culture as much an expression of status-oriented conspicuous behavior as a tension-release mechanism, that feeds back the desire to keep going in the domain where the real action is: innovation and career making at the workplace.

These cultural trends certainly vary according to the social position of the workers. An immigrant woman in an electronics assembly line will hardly be driven by the desire for innovation. Yet, the culture described is the dominant culture in the area, for two main reasons. On the one hand, there has been and continues to be a constant social upgrading of Silicon Valley, since unskilled workers and traditional ethnic minorities find fewer and fewer jobs and residences in such highly valued space, being pushed out to the outlying areas, while many of their jobs are either offshored or automated. So the dominant occupational group in the area is formed by professionals, managers, engineers, and technicians. On the other hand, even for blue-collar workers, there is a pervasive cultural influence coming out from this drive for technological poise and easy material wealth, that frames their behavior, either as a model to imitate or as an ideology to reject on the basis of their daily experience.

Thus, the values that we have outlined seemed to form the dominant culture of Silicon Valley, in the traditional sociological sense. Such values have been critical in keeping alive the innovation capacity of the area, because so much of that capacity is linked to a decentralized, entrepreneurially-based process of innovation, that combines individual ambition and organizational support of work-related social networks. This is why Silicon Valley outgrew its origins to become a self-sustaining milieu of innovation.[30] While doing so it had also to confront the reality of the problems generated by economic growth without social control.

THE QUALITY OF LIFE IN SILICON VALLEY

The experience of Silicon Valley is characterized by the stark contrast between the promises of high technology for the quality of life and the disruptive social and environmental effects produced by fast-track development on the area, on its residents, and on the industry's workers. There is, as Lenny Siegel and John Markoff wrote, a "dark side of the chip."[31] This is an important matter, because it could provide a warning sign for other areas in the world considering a similar growth process. It is also analytically relevant in order to evaluate the potential impact of the deterioration of the quality of life on the capacity of an area to keep attracting high-technology firms.

Indeed, the transformation of the urban structure of Santa Clara County under the impact of rapid industrialization in the 1950–90 period is one of the most striking examples of the contradictions between individual economic affluence and collective environmental deterioration.[32] The intensity of the process of growth put enormous pressure on scarce land – for industrial development, housing, urban services, transportation, and open space. Land prices and housing prices skyrocketed, making real estate very attractive, then adding speculative pressures to functional demands.[33] The supposedly clean industry caused serious chemical pollution, some of it stemming right from Stanford Industrial Park, contaminating water wells in many areas, including upper-middle-class areas, to the point of becoming a serious health hazard.[34] Residents in the area reacted against new development, trying to preserve the quality that had originally attracted them to the area, while raising the prices of their property. No-growth, environmental movements mushroomed, putting pressure on local governments and conflicting with the interests of future industrial development in the area.

Sharp spatial segregation came increasingly to characterize Silicon Valley. In the first stages of development, as studied by Saxenian in her first work on the area,[35] the segregation pattern differentiated four main areas: the Western foothills, with exclusive, upper-income residential communities; the North County, immediately to the south of Palo Alto, core of the high-tech industrial belt, mixing companies and middle-level residential areas; the San Jose area, for a long time the main residence for the mass of semi-skilled workers, increasingly becoming the urban support for business services; and the most distant areas, in the south of the County, where newcomers concentrated to knock on the door of the promised land.

In the 1980s, the endless expansion of the industrial and business services concentration in the Valley led to an upgrading of the activities located in the North County area, to the emergence of San Jose as a truly directional center, and to the expansion of the areas of industrial location of the Silicon Valley complex much further to the south, as far as Gilroy, and to the east of the Bay, into Alameda County beyond Fremont. This highly segregated structure, both in terms of activity and of residence, led to worse than average transportation problems, with huge traffic jams becoming an important part of the Silicon Valley lifestyle. It is not unusual to observe a young engineer driving a $50,000 luxury car with one hand while working on his portable computer with the other, taking advantage of the fact that average speeds at certain critical times do not exceed 10 m.p.h. Thus, traffic jams, increasingly higher residential density, and chemical pollution have brought significant deterioration to the environmental quality of the area. Most of the famous orchards of the Silicon Valley mythology are gone. The surface of open space per person in the City of San Jose is about one-third of that of New York City. Crime is a major problem, as in all large metropolitan areas in the United States.

And yet, throughout the 1980s, talented engineers and innovative companies from all over the world continued to settle in the Valley, as if its magnetism was endless, disregarding the forecasts of the Japanese domination in the electronics race. In fact, Japanese companies and Japanese capital are now among the major players in the industrial structure of Silicon Valley, either directly (Amdhal, owned by Fujitsu) or indirectly (substantial agreements between Japanese companies and the founding firms of the Valley, such as Intel, Apple, etc.). Thus, Silicon Valley sees its technological preeminence fully recognized, even under the new conditions of the international economy characterized by the decline of American domination in microelectronics. The obvious and rapid deterioration of the very quality of life that was an important factor in the origin of the area as a technological center, has not hampered the vitality of its economy or its capacity to innovate.

This is because milieux of innovation are a goal in themselves. People do not live in them because of the quality of their life or the beauty of their nature: quality of life is a highly subjective attribute, and many areas in the world are of startling beauty without having much chance to become technological or industrial centers. If young business talents continue to overcrowd the already overcrowded, unpleasant areas of Central Tokyo or Manhattan, it is not to enjoy the rarity of the singing of a surviving bird. It is to be part of, and be rewarded by, the world's financial centers. If film makers and music composers spend their lives on the Los Angeles freeways, it is not to catch a last ray of the sun through the ultimate toxic smog of Southern California; it is to be in the networks of the milieu generated from Hollywood 60 years ago. Similarly, the attraction that Silicon Valley continues to exercise over the high-technology researchers and entrepreneurs of the entire world relies on the simple and fundamental fact of being the depository of the most advanced knowledge in electronics and on its capacity to generate the next generation of such knowledge by processing the flows of information through its social networks and professional organizations. Silicon Valley's fate is to live up to its own historic role as a milieu of innovation of the latest industrial revolution – whatever the consequences for its land and for its people.

IS THERE A SILICON VALLEY MODEL?

Yes, there is. Not in the sense of a general formula that could and should be replicated in any other context, regardless of the economic, technological, geographic, or institutional characteristics of each region. But there is a model in the sense that we can identify, on the basis of our analysis, the elements that underlay the formation of a leading technological milieu, as well as the forms of their combination and the sequence of their development.

As we have presented in some detail elsewhere[36] in the formation of the milieu of technological innovation in Silicon Valley there concurred a number of functional preconditions as well as some key structural elements. In addition, the dynamics of the milieu itself consolidated its development. At the origin of the milieu of innovation was the historic and geographic coincidence in Santa Clara County, in the 1950s – at the very dawn of the new production system – of the three major production factors of the new informational age:

1 The new raw material, that is scientific knowledge and advanced technological information in electronics, generated in and diffused from Stanford University (since Shockley himself, while creating his own company, was also recruited to the Stanford Faculty).
2 High-risk capital provided either directly by venture capital investments or indirectly by the guaranty of military markets for still-untested devices, thus making it possible for the new companies to obtain finance on the basis of their assured earnings, regardless of their ability to succeed in their programs. The Federal Government's support through military markets and through tax provisions for small businesses was then critical for the formation of Silicon Valley. However, the reason why the Defense Department supported the effort in the critical take-off stage was because of its belief in the excellence of the technology that the companies could develop. Thus, the availability of applicable science and technology is indeed the primary factor in the development of a milieu of innovation.
3 Availability of highly-skilled scientific and technical labor in the area, from the strong electrical engineering programs of the Bay Area universities (Stanford and Berkeley, at first; San Jose State and Santa Clara, later). While Silicon Valley went on attracting talent worldwide from the 1960s, the first stages of the milieu benefited from the pool of good engineers graduating in the area, with access to support systems in the universities themselves. Thus, universities played a double role: first, particularly Stanford, as sources of new raw material: scientific-technological knowledge; and second, as providers of highly-skilled labor before the milieu could generate its own labor market.

But the clustering of these three fundamental production factors in Santa Clara County, and their articulation in a deliberate development project, were not purely accidental, unless we consider accidental the rise of an entrepreneurial project. The whole project was conceived and implemented by an institutional entrepreneur, Stanford University, under the personal impulse of its visionary Dean of Electrical Engineering, Frederick Terman. The formation of the Stanford Industrial Park, and the spin-offs from the firm founded by Shockley after his joining of the Stanford Faculty, were the material matrix out of which Silicon Valley developed.

However, the vitality and resilience of Silicon Valley over time, and the achievement of its level of technological excellence, were only possible because the Valley itself created social networks among its engineers, managers, and entrepreneurs, generating a creative synergy that transformed the drive for business competition into the desire to cooperate for technological innovation. These networks were constructed on three interrelated foundations: a work-oriented culture that valued technological genius and daring entrepreneurialism; professional organizations that sustained the interests of the Valley's electronics industry and pleaded its cause; and the territorial concentration of work, residence, and leisure that became all-embracing of its own values and interests, while excluding and segregating other social groups and economic activities.

When the Valley reached the age of maturity and the leading world corporations felt the need to be present there, they did so by incorporating themselves into the Valley's specific social organizations, whose vitality as informal networks outlasted the company affiliations of their members. Thus, the entrepreneurial and research-oriented culture, and the collective entrepreneurship of Silicon Valley's work-based, bar-reinforced, social networks, became fundamental elements for the existence and development of the milieu of innovation, just as much as the structural factors of production that made its formation possible in the first place. The interplay between the structural transformation of technology and economy, the new factors of production in the informational age, and the social, cultural, and institutional conditions of new entrepreneurship, seem to provide a necessary and sufficient explanation of the why, how, when, and where, of Silicon Valley.

As for its replicability, it all depends on the ability of firms and governments to understand the lessons of the experience, both positive and negative, in terms sufficiently analytical to be translated in different contexts; but with different strategies, and with different actors. For the very existence of Silicon Valley has changed for ever the spatial division of labor in high-technology research and production, so ironically precluding the direct replication of its own experience.

3

BOSTON'S HIGHWAY 128: HIGH-TECHNOLOGY REINDUSTRIALIZATION

If there is a success story of an old industrial region regenerating its economy from the ashes of its industrial past to become a leading high-technology complex, it must surely be Massachusetts – at least, until the late 1980s. In fact, Greater Boston has gone through at least two waves of reindustrialization in the last 40 years, after losing its traditional industries, mainly in textiles and apparel, during the 1930s and 1940s. The first round took place in the 1950s and early 1960s, and was directly linked to military and space programs that concentrated research and manufacturing in the state, mainly in precision instruments, avionics, missiles, and electrical machinery. But the post-Vietnam War reduction in military spending sent the area into another recession, producing an 11 percent unemployment rate by 1975.

Then, in the 1975–85 decade, there occurred one of the most remarkable processes of reindustrialization in American economic history: while in the 1968–75 period Greater Boston lost 252,000 manufacturing jobs, during the years 1975–80 it gained 225,000 new manufacturing jobs, most of them in high-technology industries. Most of the new firms located along Highway 128, the original suburban beltway of Boston, completed in 1951, which links 20 towns, most of them hubs of manufacturing and service activities from the old industrial era. As growth continued unabated for most of the 1980s, new industrial location took place towards the west and northwest of the state, around Highway 495, Boston's new outer beltway, joining the pioneer location of Wang Laboratories at Lowell, and spilling over the state line into New Hampshire (Figure 3.1). In 1980 there were in the Boston area about 900 high-technology manufacturing establishments, employing over 250,000 workers, to which should be added another 700 firms in consulting and technological services, making Boston the third-largest high-technology complex in the United States, after the Southern California multinuclear technopole, and Silicon Valley.

The renaissance of the town of Lowell, the oldest textile industrial city in New England, well illustrates this remarkable turnaround of the Massachusetts economy. A rural area in 1826, Lowell became a textile manufacturing

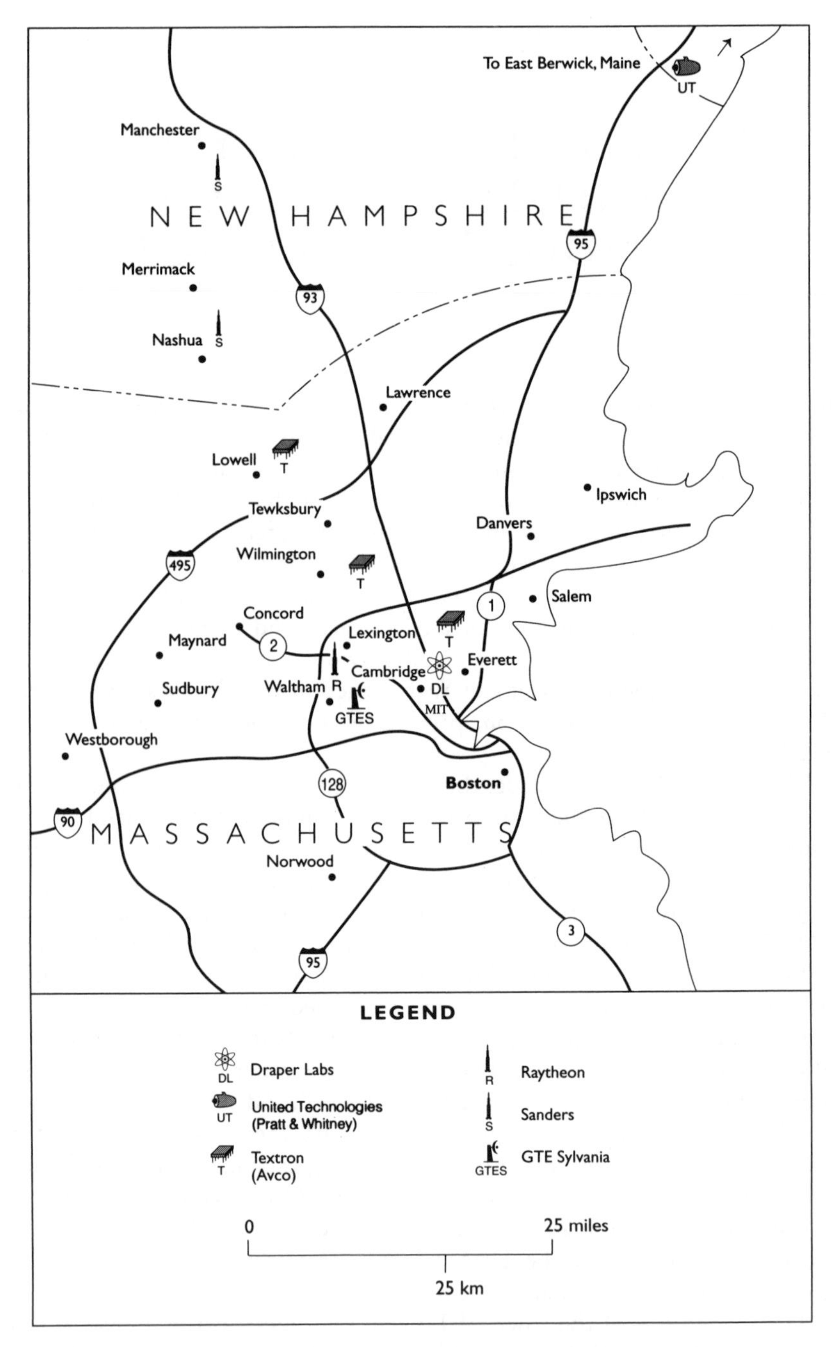

Figure 3.1 Highway 128: general location
Source: Markusen *et al.*, 1991

powerhouse from 1850 onwards, reaching a population of 112,000 in 1920, with 40 percent of its workers employed in the textile industry. But quite suddenly, in the mid-1920s, New England's textile industry shifted to the American South or went bankrupt. Between 1924 and 1932, Lowell lost 50 percent of its manufacturing jobs; it then stagnated until the 1960s.

Then, in the 1960s, a new start-up company – originally founded in Boston in 1951 by a Chinese Harvard computer sciences graduate, An Wang – established its headquarters in Lowell to take advantage of empty factory buildings, local incentives, and an experienced industrial labor force ready to settle for relatively low wages: Lowell became the center of what was for some time one of the most successful computer companies in the world, Wang Laboratories. Other companies followed the example, and during the 1970s Lowell's computer manufacturing employment skyrocketed. Electrical and electronic equipment grew at 9 percent per year in the period 1970–6, and the computer industry simply exploded in the late 1970s, growing at an annual 43 percent rate from 1976 to 1982. The computer industry was responsible for 70 percent of the growth in high-technology manufacturing, and for 45 percent of total new jobs growth during the years 1976–82. In 1980, the Lowell area had become again an industrial city, with a population of 227,000 people, and with manufacturing jobs accounting for 39 percent of total employment, in comparison with 21 percent for the entire United States; Lowell had successfully reindustrialized.

The core of Greater Boston's new industrial base is the computer industry. And the growth of this industry in Boston was linked, in the 1970s, to the growth of one particular segment of the industry: minicomputers. Some of the largest and most innovative companies in the world were born in the Boston area in the 1950s, 1960s and 1970s, and remained there: Digital Equipment Corporation (the second producer of minicomputers in the world, immediately after IBM), Data General, Wang Laboratories, Prime Computer, Computervision.

Most of these firms were new, start-up companies, created by engineers and scientists who were graduates or faculty members of MIT (less often, Harvard) or from other electronics firms, in a process of spin-off that parallels the Silicon Valley story. Indeed, a study showed that during the 1960s 175 new Massachusetts firms were created by former employees of MIT's research laboratories. The same study showed that another 39 firms spun off from an older major electronics firm, Raytheon, itself an early spin-off from MIT in the 1930s. Thus, the presence of outstanding research universities, and especially of MIT – to be precise, in Cambridge, on the banks of the Charles River opposite Boston – was certainly a decisive factor in the ability of Massachusetts to reindustrialize.

Furthermore, location surveys conducted on the area's high-technology companies show that the critical factor for maintaining their location here is direct access to one of the largest concentrations of academic, scientific, and engineering talent in the world. Indeed, there are 65 universities and colleges in the Greater Boston area, some of them of top international quality, such as MIT, Harvard, Amherst, Tufts, and Brandeis, and others of respectable quality, such as Boston University, the University of Massachusetts, and Northeastern University, in addition to several elite undergraduate liberal colleges. The whole of New England contained over 800,000 undergraduate and graduate college students by the late 1980s. This factor is critical for high-technology industries, since college graduates represent about 33 percent of all employees in the computer industry. Engineers and technical workers accounted in 1980 for 9.1 percent of employed workers in Massachusetts, in contrast to 7.3 percent in New York, 8.0 percent in Illinois, or 8.9 percent in California.

Thus, apparently, Massachusetts' successful reindustrialization resulted from a unique combination of an entrepreneurial-industrial tradition and the excellence of a university system that provided the necessary new raw material, scientific knowledge and technological skills, which could generate a structural shift from smokestack industries into a high-technology and advanced services economy, in a process largely independent of government policy. Yet, as usual, the story is more complex; it demands careful examination of the historical origins of the Massachusetts scientific-industrial complex.

MIT, THE WAR MACHINE, AND THE NEW ENTREPRENEURS

Clearly, at the heart of Greater Boston's high-technology development lies a strong foundation of scientific and technological excellence. Other factors – good communications and telecommunications, cultural amenities, historic heritage, beautiful landscape, and the like – are certainly not unique to Boston. The presence of venture capital firms, accounting for the fourth-largest concentration of such firms in America, after New York, San Francisco, and Chicago, seems to be a consequence of high technology development rather than a causal factor. The formation and subsequent growth of this scientific-technological complex certainly owe a great deal to its strong tradition in higher education and academic research. But the shaping of this academic basis to the service of leading-edge industrial technologies, and the considerable financial and human resources that were concentrated in such an effort, are mainly linked to one institution, MIT, and to a particular process: the technological shift in warfare, first during World War Two, later during the Cold War. We need to trace back the story from its origins.[1]

In the 1930s, MIT had two features that, together, made it different from all other elite academic institutions on the East Coast: it had the oldest and most distinguished electrical engineering department in America, founded as early as 1882; and, because it had much less money than Ivy League universities such as Harvard or Princeton, it was much more open to conduct contract research with the Government or with private firms. Indeed, such contractual cooperation was (and still is) an established policy at MIT. It also spearheaded the movement towards industrial spin-offs: in 1920, an associate professor in electrical engineering at MIT, Vannevar Bush, created a company, Raytheon, to produce thermostatic controls and vacuum tubes.

Both Bush and his company would prove decisive for the future of high-technology in Massachusetts, with the help of World War Two and of the Defense Department. In 1940, Bush became director of the Federal National Research and Defense Committee, and in this position was fully conscious of the decisive military potential of radar, just invented in Britain. To prevent the capture of such critical knowledge by the Germans in the dark hours of Britain's heroic resistance to the Nazis, Bush convinced both the American and British Governments to bring the British radar research team to the United States to continue secret work on the device, together with Professor Bowles of MIT. After laboratory space was arranged at MIT at a few hours' notice, the team went on to create MIT's Radiation Laboratory, so named to mislead the Germans about the nature of the research being conducted there. Radar and its applications came out of the Radiation Laboratory, later transformed into the Electronics Research Laboratory, one of the most distinguished institutions in the field.

Bowles also worked with Raytheon, still a modest company, and he obtained for his company the contract to manufacture the necessary equipment, sending Raytheon into explosive growth: by 1945 Raytheon, located in Waltham, Massachusetts, had increased its labor force 40 times, to become a large company, producing 80 percent of the world's magnetrons. In the 1950s, Raytheon became a major industrial force in the field of rockets and missiles.

Other advanced laboratories were founded at MIT in the 1930s and 1940s, on the basis of major military research contracts, among them: the Instrumentation Laboratory, today an independent organization; Draper Labs; and the Lincoln Laboratory, established after the war to contract research for the Air Force on radar and computer technology. Harvard also played a role, albeit more limited, in the establishment of military-oriented, electronics research programs, particularly on the basis of the Harvard Countermeasures Laboratory, established in parallel to the Radiation Laboratory.

But perhaps the most direct connection between the MIT-induced wartime programs, and today's high-technology industry in Boston, was the formation and development of a computer science capability at MIT, following the work of Jay W. Forrester, who arrived at MIT from his native Nebraska in 1939, and worked during the war at MIT's Servomechanisms

Laboratory. To solve the complex simulation problems he had to resolve to help the demands of aviation, he decided that he needed something that then did not exist: a computer. He visited John von Neumann at the University of Pennsylvania, who was building the machine that would become the first computer, the ENIAC, which became operational in February 1946 – a remote application of the ideas published in England in 1937 by the mysterious genius at the origin of this scientific adventure, Alan Turing.

Forrester was disappointed with Neumann's machine: it was too slow and too unreliable. He started his own project, the Whirlwind Computer Project, and sold it to the Pentagon as the core of a system to build an "electronic radar fence," the "Star Wars" of the 1950s: the Semi-Automatic Ground Environment, or SAGE project. Forrester selected for the work a company that – unlike Raytheon, Sylvania, and Remington Rand, the earliest major electronics firms – was not yet established in electronics. This new company was IBM, and it was chosen by pure chance. In June 1952, when Forrester was looking for a commercial manufacturer to build his iron core memory for the SAGE project, he met at the Second Joint Computer Conference an IBM engineer, John McPherson, who seized the opportunity, convincing IBM's President Tom Watson to meet President Truman to offer the services of IBM. Having secured the contract, IBM started moving from the punch-card business to computer manufacture. On the basis of the SAGE contract, IBM hired 8,000 engineers and workers, signaling its real head-start in computer manufacturing with its Model 650, delivered in December 1954.

IBM stayed aloof in its northern New York semi-rural estate, mirroring the isolation of the other technological giant, AT&T, in its New Jersey suburb of Murray Hills. But the seeds of innovation, which were sown in the Whirlwind program and in the SAGE project, were germinated in Massachusetts. Jay Forrester, who did not entirely trust IBM, sent one of his graduate students, Kenneth Olson, to supervise the company's work at Poughkeepsie. Olson deeply disliked IBM's corporate style. So, once his work was finished, he decided to start his own computer company in an abandoned factory in Maynard, Massachusetts (near the later Route 495) in 1957. The name of the company was Digital Equipment Corporation. A new industry grew out of new knowledge. But this new knowledge was nurtured by Pentagon financial and institutional support.

In fact, the continuing support of the defense establishment, both for MIT and Harvard work and for MIT-originated companies, was based on the ties of trust, both personal and scientific, that linked the MIT–Harvard network with the commanding heights of the US Government for three decades. At the end of the 1930s, President Roosevelt appointed Vannevar Bush as his science adviser. In the 1940s, Truman's science advisers were James Killian from MIT and George Kistiakowsky from Harvard. In 1960, Kennedy's science adviser was Jerome Wiesner. Johnson made a small variation: he appointed Don Hornig from Princeton, but Hornig was a Harvard college graduate and a Harvard Ph.D. Nixon appointed as science adviser Lee Du Bridge, former director of MIT's Radiation Laboratory. Ford appointed Guyford Stever from MIT, and Carter named Frank Carter, also from MIT. Thus, a long string of science presidential appointees from MIT, and to a lesser extent from Harvard, marks the critical importance of the connection between the centers of decision of the Federal Government, particularly in technology-based defense programs, and the technological potential of MIT. It was thus no surprise when in the 1980s the "Star Wars" Program reactivated the direct linkage between defense and advanced research, MIT again received a substantial proportion of the new contracts, in spite of the established dominance of California as the main high-technology defense-oriented complex.

To sum up, the causal chain of events underlying the development of Greater Boston's high-technology complex operated thus: MIT (and to a lesser extent Harvard) became the hub of advanced electronics research in the 1940s and 1950s, with considerable support through funds and orders from the Defense Department. MIT's Faculty and graduates used their advanced knowledge in new technologies, as well as their excellent personal contacts with the military establishment, to start companies that prospered rapidly because of lack of technological competition in the new industrial complex. These companies reproduced the spin-off process, giving birth to dozens of new companies that clustered in an industrial-technological milieu, developing agglomeration economies, and supported by the high quality of labor in the region, which arose from its educational basis and its industrial tradition in skilled manufacturing.

Yet, really to understand the formation of the Boston high-technology complex, we must add another layer of complexity, by analyzing the evolution of its industrial composition at different stages of its development. This new round of analysis will in fact prove critical for our ability to judge the experience and thus evaluate the region's future prospects.

THE FOURTH INDUSTRIAL WAVE: THE TUNNEL AT THE END OF THE LIGHT

While these elements are all key ingredients of high-technology growth in Massachusetts, their relative importance differed quite markedly at various stages of the process.

In the 1940s and 1950s the beginnings of an electronics-based defense industry took place around the old industrial establishments created in the 1920s, Raytheon and Sylvania, and the much older establishment of aircraft engine manufacturing plants by General Electric and Pratt & Whitney in Lynn, an old industrial town north of Boston. But, in spite of the scientific support of MIT, this older defense-oriented complex could hardly survive the decline in defense spending at the end of the 1950s, after the Korean War, and in the early 1970s, when the Vietnam buildup was brought under control.

In particular, the oldest companies were unable to adapt from vacuum-tube technology to transistor technology during the critical evolution of the electronics industry in the late 1950s. Indeed, Raytheon rejected Shockley's offer to cooperate with the company, prompting Shockley to go back to his native Palo Alto to found Shockley Semiconductors, the ancestor of most of Silicon Valley's semiconductors companies. The failure of the old electronics companies to enter the microelectronics field explains why, in spite of its early start and superior scientific base, Massachusetts lost its leadership to California in electronics manufacturing. Thus, the first reindustrialization in the 1950s, tightly linked to the defense industry both in electronics and engines, did not pull the region out of its structural decline, because of the rapid bureaucratization of large companies after their innovative stage, and because of excessive reliance on cozy arrangements with the defense establishment.

The third wave of industrialization in Massachusetts, based on computers from 1975 onwards, was thus quite independent of the old 1950s electronics base.[2] It was started by new companies, the great majority of which were created after 1960, with the addition of the most important of all, DEC, founded in 1957. The computer industry did have an original linkage with military programs, because most of the new entrepreneurs obtained their knowledge in research projects linked to MIT or to other firms generally funded by defense spending. Yet the products of the new companies diversified quite rapidly; by the boom decade, 1975–85, they were mainly aimed at the civilian market. It was the invention of the minicomputer and the introduction of computers into the office (with the workstation concept developed by Wang) that propelled the new high-technology complex in Massachusetts in the late 1970s.

However, the Reagan administration buildup in the 1980s again redirected Massachusetts' high-technology industry toward military programs: New England became the region that received most defense spending per capita in the entire United States during the 1980s. The emphasis of "Star Wars" on software and artificial intelligence, one of the strongest scientific fields at MIT as well as in the technological services firms of Massachusetts, created huge, instant, highly-profitable markets for these firms, precisely in the critical 1984–6 period when the world's computer industry went into a downturn. Seizing the opportunity, Route 128's computer companies became increasingly dependent on military markets, reversing the trend that had begun the 1970s.

Thus, what appears at first as a continuing process of high-technology development linked to defense markets, in fact conceals three stages: the World War Two and Cold War buildup of an industrial base, around large firms operating as defense contractors; the spin-off of entrepreneurial firms in the 1960s and 1970s, creating a new, civilian computer industry out of the military-oriented research programs; and the remilitarization of the computer industry, particularly of its technologically advanced components, during the 1980s, under the combined pressures of a slowdown in the world computer market and stepped-up demand from the new Pentagon technological frontier.

The vulnerability of such excessive dependence on military markets became evident during the 1988–91 period, when the Massachusetts miracle came to an abrupt halt: with the reversal of defense spending, because of budgetary constraints and the end of the Cold War, high-technology industries laid off some 60,000 workers in three years, and unemployment surged in the state. But the relative weakness of the Massachusetts high-technology complex may have deeper roots than the standard explanation of excessive reliance on military markets.

One reason may be technological. The region is much less diversified in its industrial high-technology structure than Silicon Valley or other high-technology areas. Computer making accounts for the bulk of the activity. And within computers, one particular kind of computers, minicomputers, dominates the picture. Technological change in microelectronics in the late 1980s dramatically increased the power of microcomputers, and advances in telecommunications, in software, and in the protocols to link up computers, have made possible the shift from minicomputers to networks of microcomputers working together as a unit. Thus, companies heavily specializing in office-oriented minicomputers, such as Wang, were substantially hurt by the new developments, as was Digital, in spite of its greater diversification. The industrial structure of the region must thus adapt to the new conditions of technology and to the growing competition in the world market, as Japan, Europe, and the Asian newly industrializing countries (NICs) enter the computer market.

Given that fact, a second potential reason for the new industrial crisis in Massachusetts appears to stem from the relative incapacity of its industrial structure to adapt to the new technologies and to the new market environment. In her latest work, AnnaLee Saxenian emphasizes the lack of flexibility of the Route 128 firms, in spite of their entrepreneurial origin. Three elements seem to explain such industrial rigidity: first, many of the firms were in fact spin-offs of older industrial firms, such as Raytheon, and their founders brought with them their old-fashioned corporate culture; second, the industry is ruled by professional associations, such as the Massachusetts High Technology Council, that enforce an industrial discipline so as to exercise their lobby power, both in the State Government and in Washington; finally, the renewed reliance on defense contracts in the 1980s diminished the entrepreneurial skills of the firms, removing them from the fast-changing networks of the world's computer markets. So the future of the Greater Boston high-technology complex looks gloomier than that of other similar areas in the United States. Ultimately, its future must depend on the ability of its firms to make the critical organizational and technological transition to the peace economy and to the world market.

Thus, reindustrialization is a magic word with little meaning unless it takes place within a particular historical, technological, and institutional context. The Greater Boston area has undergone not one, not two, but three reindustrialization processes in the last 50 years, each one of them characterized by a particular mix of relationships between research institutions, government markets, and industrial structure. What remains a constant factor, throughout the process of repeated recovery from the depth of crisis, is the state's commitment to excellence in education and research, which it inherited from its founding fathers. It remains to be seen whether that commitment will bounce Massachusetts back for yet a fourth time.

4

SCIENCE CITY BLUES: INNOVATION BY DESIGN?

Science cities are new settlements, generally planned and built by governments, and aimed at generating scientific excellence and synergistic research activities, by concentrating a critical mass of research organizations and scientists within a high-quality urban space. In some cases, they are intended to link up with industrial firms or to generate commercial spin-offs; in other instances, not so. But what characterizes science cities, in contrast with other types of technopoles, is their focus on science and research, independent of their impact on their immediate productive environment. They are generally conceived as supports to national scientific development, considered a positive aim in its own right, in the hope that better scientific research will progressively percolate through the entire economy and the whole social fabric. They are also often presented as tools of regional development, intended to assist the decentralization of scientific research, with all the prestige that involves, to the national periphery or, failing that, the metropolitan periphery.

Underlying the experience of science cities is the notion of the campus model as the spatial expression of research, innovation, and high learning. To build a community of researchers and scholars, isolated from the rest of the society – or at least from its vibrant urban centers – is an old, well-entrenched idea; in western societies, it goes back to the medieval tradition of monasteries as islands of culture and civilization in the midst of an ocean of barbarism. The construction of such privileged, secluded spaces is intended to signify a certain distance from the day-to-day conflicts and petty interests of society at large, potentially enabling scientists and scholars to pursue their endeavors, both detached from – and independent of – mundane material concerns. Simultaneously, the internal closure of the space is supposed to spur the cohesion of intellectual networks that will support the emergence, consolidation, and reproduction of a scientific milieu, with its own set of values and mechanisms to promote the collective advancement of scientific inquiry.

Yet paradoxically, science city projects also tend to be linked to the all-powerful will of a Prince (in modern terms, an autocrat or technocrat) with the power to create *ex nihilo* a new site for science, which will force into privileged seclusion some of the best research talent of the country. Their symbolic value is as important as their functional potential, since they provide the material proof of the State's commitment to science and technology, while their spatial concentration and institutional coordination make visible the national scientific resources.

More often than not, private firms and institutions are eventually called upon to link up with public research centers and universities in the establishment of science cities. But the goal remains the emphasis on research *per se*, with private and public ventures cooperating in a common interest in the advancement of science. The more a society doubts its capacity to generate scientific excellence through the spontaneous vitality of its institutions, the more it tries to leapfrog stages of development by concentrating scientific resources in space and time, so as simultaneously to increase synergy, to gain visibility, and to generate a culture of academic excellence, thus to act as a precondition for economic dynamism and political power in the new information-based society.

Given that in most modern societies scientific institutions are highly dependent on governments, their spatial location is often easier to manipulate than the investment decisions of private corporations. Thus governments, vowing to decentralize major functions from major cities and/or from traditionally dominant regions, are always tempted to exile research institutions to new areas, forcing them into the pioneering role of disseminating knowledge in the virgin lands – that is to say the uninformed territories – of the information economy.

The results are rich, diverse, and indeed contradictory. More often than not, they substantially depart from the intentions of their would-be creators; yet they always introduce new dynamics, both in their regional environment and in their internal working. To approach this phenomenon we have found it useful to start by analyzing the creation of a science city under the conditions of total state control over research, industry, and spatial location – conditions that should favor the complete articulation of all the necessary ingredients of a synergy-generating science city. This case is Akademgorodok, near Novosibirsk, personally designed by Khrushchev in the late 1950s as part of his grand vision to develop Siberia into the future powerhouse of communism. We will then analyze another major project resulting from a direct presidential decision to foster science and regional decentralization by building a science town, as it happened in the President's native province: the Taedok Science Town built from the mid-1970s onward near Taejon, South Korea. Moving from authoritarian political decisions to the implementation of technocratic logic, we will then present the process of building and developing Tsukuba Science City, near Tokyo, perhaps the best-known science city experiment in the world. Finally, we will discuss the current transformation of the concept within Japan in the 1980s, by studying Kansai

Science City, between Kyoto, Osaka, and Nara: a new urban form that brings together public and private organizations from the outset, and that moves away from the old locality-based concept of scientific city to introduce a new model of a linked, multinuclear development, apparently characteristic of the space of flows within the informational economy.

Throughout the itinerary of this intellectual-geographic journey, we shall focus on three fundamental questions. First, what is the added scientific value to be obtained by spatially concentrating resources and institutions? Second, is it possible to link scientific research, technological applications, and industrial and commercial productivity, given a sharp spatial separation of each function? And finally, what is the impact, if any, of pure science cities on the processes of regional development and economic decentralization? By careful reconstruction of these little-known experiences, we may be able to gain tentative answers to these very old, yet still unresolved, questions.

THE SIBERIAN DREAM: AKADEMGORODOK[1]

The creation of one of the largest and most advanced scientific complexes in the world, in the middle of the Siberian birch forest, has been one of the boldest experiments to date in using a new town as a direct instrument of techno-economic development. In 1957, advised by a leading Russian mathematician, Lavrentiev, Khrushchev decided on the foundation of Akademgorodok, on the banks of the Ob Sea (a gigantic dammed lake), 15 miles (25 kilometers) from Novosibirsk, the main industrial center of Siberia. The first buildings – research institutes, a new university, and residential areas provided with urban amenities – were completed in 1964. An elite university, and 20 large research institutes of the Siberian Branch of the USSR Academy of Sciences, form the workplace for about 7,500 scientists, 3,500 students, and 1,500 university staff members, besides several thousand workers and technicians employed by the research institutes. In 1990, about 70,000 residents lived in Akademgorodok, forming part of a wider administrative unit, the "Soviet District" with a population of 125,000 people, which in turn is one of ten districts making up the even larger Novosibirsk region, totalling 1,500,000 (Figure 4.1).

Akademgorodok has indeed become a city, with a pleasant landscape, large avenues cutting across the forest, and concrete cubicles that offer a somewhat better physical appearance than the replicas of Stalinist architecture that dot every city in the Soviet Union. Major scientific advances have taken place in the research institutes, and the university has trained Siberia's best scientists. Akademgorodok has housed some of the best researchers in the entire Soviet Union, including several top Gorbachev advisers in the first period of *perestroika*. There is no doubt that Akademgorodok was infused, at least for a time, with unusual vitality and scientific creativity. But to

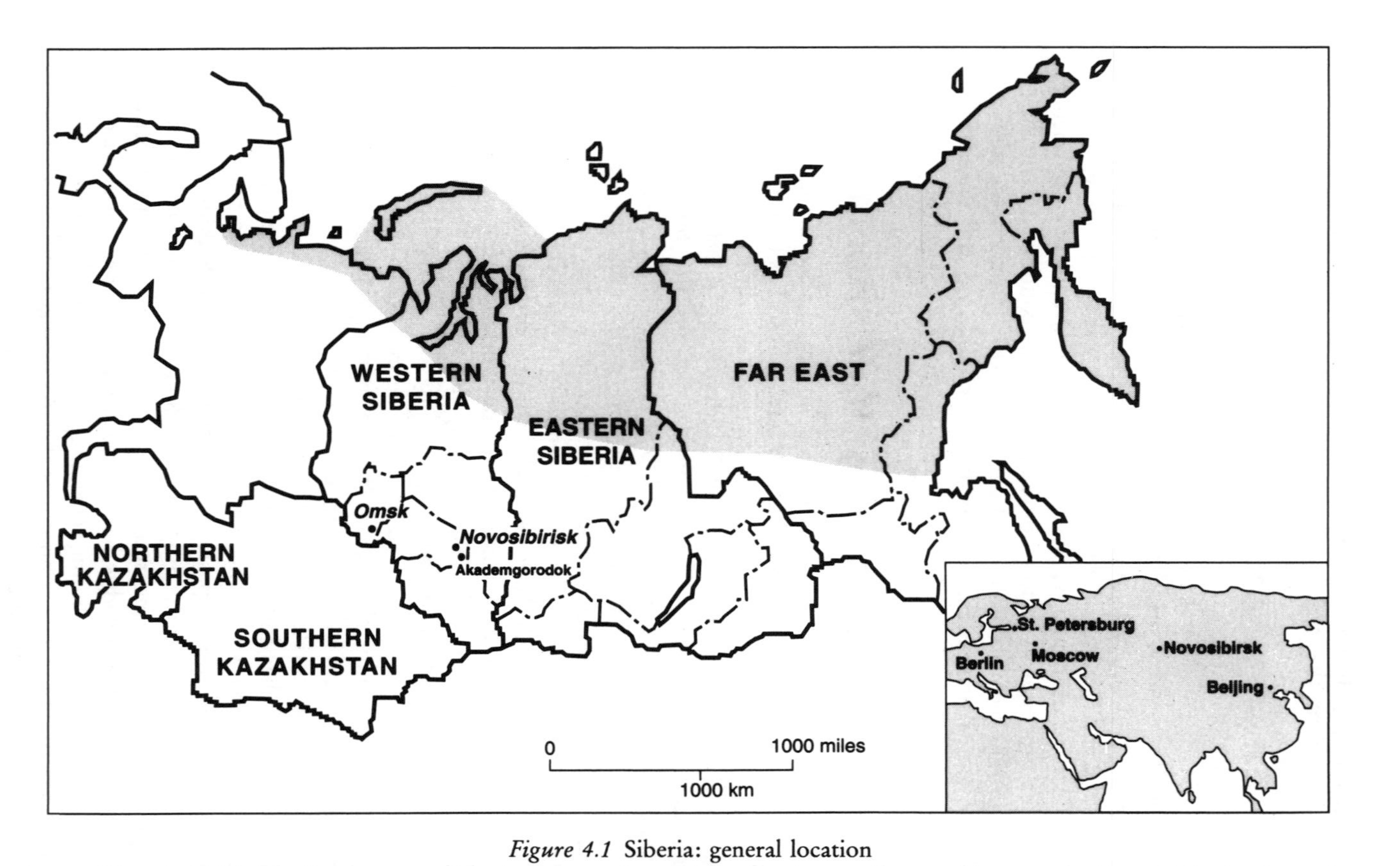

Figure 4.1 Siberia: general location

Source: Diens, L. (1982) "The Development of Siberian Regions: Economic Profiles, Income Flows, and Strategies for Growth," *Soviet Geography: Review and Translation*, XXIII (4), April

Note: The broken lines represent *oblast* boundaries

answer the broader question concerning its role in Siberian regional development, as well as its contribution to technological innovation in the Soviet Union, we need to examine its history within a larger framework: of the process of industrialization of Siberia, and the contradictory evolution of the Soviet economy, over the quarter-century from 1965 to 1990.

The origins

Akademgorodok was born out of the Khrushchevian dream: to achieve communism, through science and technology, by the end of the twentieth century. As the story runs, during his trip to the United States in 1957, Khrushchev was impressed by the quality of American research universities, and by the role of campus life in helping to create a scientific milieu. Sharing his thoughts with one of his scientific advisers, the leading mathematician Academician Lavrentiev, they came to conceive the project of a new scientific city, that would concentrate some of the best talent in the Soviet Union and would link their research with industrial enterprises, to become a model for the modernization of Soviet industry.

This accorded well with Khrushchev's own vision of developing Siberia and the Far East, to take advantage of their untapped wealth in natural resources and to populate and colonize that entire half of the Soviet territory that remained sparsely inhabited. Besides, this pioneer spirit, in the views of Khrushchev and his advisers, would help shake up the bureaucratized Soviet academic institutions, giving the opportunity for rapid advancement to a new generation of scientists whose creativity was hampered by established senior academicians in the main science centers of Moscow and Leningrad. The chance to move to a new area, to start new research institutes, with greater autonomy and considerable resources, proved indeed appealing to a number of bright young scientists frustrated by bureaucracy. It proved equally attractive to other scientists, young or less young, who happened to be ideologically liberal or of Jewish ancestry, and so found themselves blocked for promotion. Thus, Lavrentiev did succeed in recruiting some very talented scientists, particularly physicists like Academician Budker, one of the leading nuclear physicists in the world. Counting on a substantial scientific input, Khrushchev in 1960 created a Siberian Branch of the Soviet Academy of Sciences, to be headquartered in the new scientific city.

The site of the city was determined by Lavrentiev himself, on the basis that it should contribute to the development of Siberia. This naturally led him to examine Novosibirsk, Siberia's largest industrial and urban centre, as the prospective location. The model of a scientific campus, relatively distant from a major city, concentrating on its work and forming a real community, was deliberately sought. Lavrentiev and his planners boarded a helicopter and toured the region. Thus they spotted a beautiful, natural area, in the middle of the birch forest, at the point where the Ob river is dammed to form a gigantic artificial lake. Furthermore, the area, although only 15 miles (25 kilometers) from Novosibirsk, was absolutely uninhabited, allowing for a new settlement in totally virgin land.

The decision to build the new town there was taken on the spot. But it turned out that there was a good reason why the area was empty. The beautiful forest, during springtime, was full of tree-leeches whose bite provokes meningitis. Because spring is the only good weather period, this circumstance effectively makes the forest useless for recreation. When Lavrentiev heard of the problem, after the decision to locate the city had been made, he remained adamant: scientists would take care of the problem, he argued, meaning that they would spray the forest with insecticides. They did. But in the process they also wiped out most of the birds, to the point that they had to stop the treatment. The insects came back *en masse* and have remained ever since. It was somehow representative of the hard realities that would attend the creation of a scientific utopia in the "academic little village" (*Akademgorodok* in Russian).

Planning the city

Construction of Akademgorodok started in 1957, and the first stage was finished in 1964. In fact, most of the city was built in the first stage, during the 1960s, while expansion was later slowed down. The planning was strictly functional: it aimed at social and functional segregation of land uses. It established a distinction between three types of zones:

1 The zone for the scientific institutes (depending on the Academy of Sciences).
2 A university area to house the new Novosibirsk State University. Adjacent to the university buildings were students' dormitories, carefully separated from other residential areas, so that students and scientists would not mix in the same spaces. In fact, they did not.
3 Residential zones, with a sharp differentiation between:
   (a) High-quality residential cottages for academicians and distinguished professors, also provided with their special shops and leisure centers.
   (b) The "upper zone," formed by apartment buildings of reasonable quality (but with apartments as small as in Moscow, about 12 square meters per person), for middle-level scientists.
   (c) The "lower zone," with buildings of lower standards, intended to house the construction workers and the manual workers of the research institutes.

In reality, while the academicians' cottages remain a secluded, exclusive world, and the students' dormitories have created a distinctive subculture, the rest of the city has become rather mixed, with everybody occupying whatever housing unit they could get, by whatever means, on arrival in the city. All the areas are scattered in the middle of the forest, separated by wide avenues that have become the main axes of connection between the sharply-differentiated zones. Thus the forest provides a natural reinforcement and a physical definition of the intended zoning.

In the original plan, there were also meeting points, sports halls, a stadium, a swimming pool, a "pioneers' house" (for youth), a musical school, a hotel, and other facilities. Most of these amenities were never built. In 1990, the main social points were a multi-service social club and meeting building, linked to the Academy of Sciences, one upscale public restaurant, and a few alcohol-free, student canteens. The only hotel in town, run by the Academy of Sciences, was forbidden to non-guests.

One of the key features of the original Lavrentiev Plan was the location of industrial enterprises close to the new city, so that contacts would be established between research and production. However, the ministries responsible for the industrial enterprises decided otherwise, since they wanted each firm to rely on its own research institutes, not on the Academy of Sciences. Thus, there is no industrial zone in Akademgorodok. Years later, however, a number of Novosibirsk firms decided to locate plants in the area surrounding Akademgorodok, to take advantage of existing land and transportation facilities. Yet these factories formed no part of the original plan; and they have no relationship whatsoever with the scientific institutes in the city.

Thus, the city is organized around the activity of the 20 research institutes of the Siberian Branch of the USSR Academy of Sciences located in Akademgorodok. The institutes are very large institutions, employing hundreds or even thousands of scientists and workers, and many of them have their own industrial shops, produce their own machinery, and take care of their own repairs, and even of their own construction work.

A critical feature of the original planning concept was to separate Akademgorodok from Novosibirsk. Thus, there is no train service, and only one road with a public bus connection – not the most convenient means of transportation when it implies waiting for perhaps 30 minutes in the open air in a temperature of minus 30°C. Cars are few, and are in any case useless in winter because there are few garages and cars cannot be left parked outdoors in these temperatures. The obvious purpose of this plan was to preserve the social integrity of Akademgorodok as a scientific community, separated from the industrial working-class city of Novosibirsk. This created great hostility on the part of the local authority (the Novosibirsk Soviet), which always saw Akademgorodok as an artificial implant, siphoning Moscow funds for the gratification of a useless intelligentsia. It responded in predictably petty ways, for instance by allocating less water than the city needed. The Academy of Sciences took charge of as many practical needs as it could, bypassing the local authorities, thus fomenting the hostility.

Such an isolationist design could prove viable only on condition that the overall housing supply would improve, thus ensuring that people could find housing close to their work. In the early 1960s, Soviet planners axiomatically assumed that by 1990, even if the country had failed to reach the communist stage, the housing question would be solved. But it was not. On the contrary: the crisis has worsened, thus becoming a total deterrent to mobility of labor. Thus Akademgorodok children, after finishing secondary or university education, must remain in their parents' apartments. They cannot find jobs in the local scientific community, since the administration only replaces losses from retirement or departures. Together with women entering the job market, the result has been a surplus of some 8,000 Akademgorodok workers who can only find work in the much larger, diversified economy of Novosibirsk. So, despite the planners' dream, Akademgorodok, like many other new towns around the world, is becoming a suburb. The difference, as compared with most suburbs, is that the planners have provided virtually no transportation to the city.

Productivity

Akademgorodok was conceived as a milieu of scientific innovation which would serve industrial development in Siberia and in the Soviet Union generally. In order to work, it would have had to fulfill the following conditions.

First, the scientists would have to be both top-quality and highly-motivated; they would have had to be engaged in constant interaction in an atmosphere of freedom and creativity. Second, linkages would need to be established between research institutes; they would need to share services and facilities, and set up complementary research programs. Third, backward and forward linkages would need to be developed between research institutes and industrial enterprises, both in Siberia and in the Soviet Union. Fourth, the entire complex would need to be opened up to the outside world, with scientific international cooperation and frequent exchanges with other academic centers in the Soviet Union. Fifth, there would need to be cross-fertilization between the research institutes and the university, both for faculty and for students. Sixth, the research institutes would need to evolve, recruiting outstanding new researchers, and keeping their senior personnel.

This set of conditions is not just our reconstruction; it is the actual list of goals assigned to the scientific complex at the start. But, 30 years later, in spite of the high quality of much of the scientific research carried on in its institutes, the record shows almost complete failure to achieve the initial objectives.

Early on, in the 1960s, Akademgorodok was able to attract many top-quality young scientists, anxious to escape the dead hand of the Moscow bureaucracy. A liberal atmosphere and a pioneer spirit prevailed. Social interaction was easy, leading to passionate discussions and a fairly lively social life. The "Integral Club," mainly attended by physicists, had seminars and talks downstairs, dancing and drinking upstairs. The university tolerated dissident activity, such as the only public demonstration against the Soviet invasion of Czechoslovakia in 1968.

But the change in the political climate, the increasingly oppressive controls of the Brezhnev era, and the bureaucratization of the Siberian Branch of the Academy of Sciences, started to erode the initial enthusiasm. Many scientists moved back to Moscow and Leningrad at the first opportunity: for instance, 100 of the 150 nuclear physicists who originally moved in left Akademgorodok after only five years. From the late 1960s onward, it became increasingly difficult to recruit first-class scientists to Siberia, despite incentives. Housing shortages reduced mobility, literally freezing people's location both in Siberia and in the European parts of the Soviet Union.

The institutes increasingly resorted to recruiting their new personnel from the graduates of Novosibirsk State University, and from the lower-level Siberian research institutes in other cities. And so – apart from a few highly visible programs, particularly in nuclear physics, in mathematics, in chemistry, and in economics – Akademgorodok became increasingly provincial and isolated. Very little exchange with Western scientists took place during the 1970s, most of it with Germany.

This isolationist pattern also became the norm for relationships between institutes. Institutes are very large organizations, with hundreds, sometimes thousands, of researchers and workers; and they became self-sufficient entities, only accountable to the Presidium of the Siberian Branch of the Academy of Sciences. Their lack of cooperation reached the point where most of them established their own computer services, instead of sharing scarce facilities and personnel. They also established their own libraries, supply systems, and housing and services for their personnel. Scientific interaction between scientists from different institutes, and even between different departments of the same institute, became extremely rare. Each researcher worked on a specific line of research controlled by a supervisor, who in turn was directly accountable to a program director. According to our interviews with researchers, even in 1990, any project that was not strictly controlled by the bureaucracy of the Academy would not receive support, thus effectively halting it. The youngest and best researchers reacted by withdrawal from the system, either moving to Moscow – where they would be equally controlled, but where living conditions were better – or by fulfilling their minimum quota requirements.[2]

There were a few exceptions to this bureaucratic dead hand – as for instance in the largest and most prestigious institute of Akademgorodok, the Nuclear Physics Institute. This has continued to perform leading-edge research in high-energy physics, because the caliber of its leading scientists compelled the Academy to respect its autonomy, including budgetary matters, and because it could sell advanced experimental machinery (such as synchotron radiation devices) to other institutes and enterprises in the Soviet Union. In fact, we personally observed this: every day, at noon, approximately 100 directors of research teams meet in a very large room, around a circular table, for about 10 minutes of business, followed by informal discussion on every aspect of their work, exchanging papers, addresses and information about research activities, or about seeking invitations for scientific meetings abroad. Discussion is open and frank; scientists treat each other as equals. But such events are rare in Akademgorodok.

Yet the main failure in the original project, 30 years later, was the almost total lack of linkages with industrial enterprises, whether in Novosibirsk, in Siberia or, for that matter, in the Soviet Union. The institutes concentrated mainly on basic research, but even their findings in this area were not diffused or linked to any industrial application, even in cases where such applications were fairly obvious. Research findings were reported to the Academy of Sciences, or published in academic journals, which made them accessible to the ministries that would communicate them to their own research institutes. These ministerial institutes were responsible for meeting the needs of the firms within their own production area, but they used research findings in a very narrow way, simply to respond to immediate requests.

The isolation of the Akademgorodok institutes from industrial firms was such that each major institute had its own industrial shop (actually a factory) in Akademgorodok, to produce the machinery it needed for its experiments. Only rarely did they share or sell their machines to other institutes. For instance, experimental radiation machinery produced in Akademgorodok was sold to Moscow centers, as well as to German centers by the Nuclear Physics Institute. Scientists from other Siberian Institutes, who refused to buy the machines, were forced to travel to Germany to use the machines for their experiments. The problem lay with the budgetary and administrative structure of the Academy of Sciences, which was entirely independent of that in the industrial ministries, and which specified a specific budget line for each territorial branch and for each institute within that branch (Figure 4.2).

The isolation from other scientific units, both in the Soviet Union and abroad, and the almost total absence of linkages with industrial applications, seems to be responsible for the striking paradox of Soviet science and technology: a very strong scientific basis has coexisted with increasing technological backwardness in key areas of research and application. This was particularly striking in the fields of informatics and microelectronics, as can be illustrated by the observation of two of the leading research institutes in Akademgorodok.

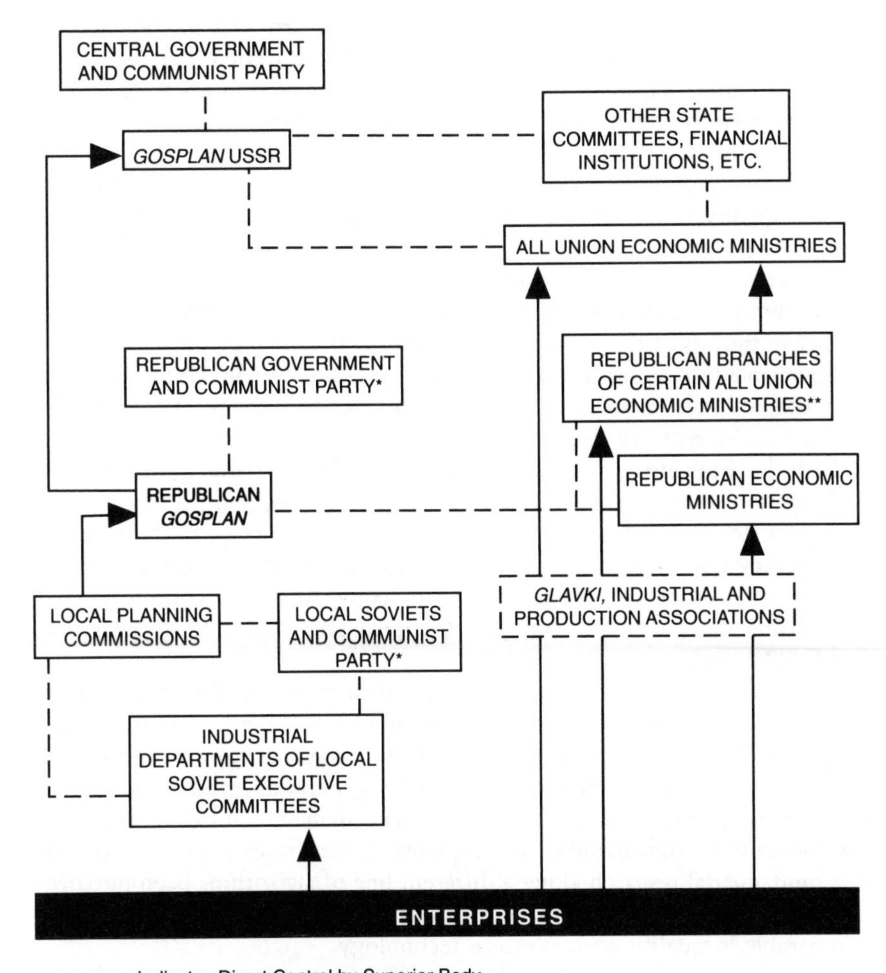

Figure 4.2 The Soviet economic planning system
Source: Pallot and Shaw, 1981

The new Institute of Informatics Systems, formed in 1990, was created as a spin-off of the Institute of Applied Mathematics and Informatics, which was the depository of a very strong research tradition in theoretical informatics, derived from leading mathematicians, such as Lavrentiev, Marchuk (the 1990 President of the USSR Academy of Sciences), and Gerschov. In 1990 the new Institute, along with nine other laboratories, was able to organize an advanced laboratory of artificial intelligence on informatics systems. Researchers in these institutes argued that their software was better than its Western equivalent; their computers were so bad that they were forced to use creativity to make inferior machines work to their maximum potential.

Why, then, are Soviet computers so bad? Part of the explanation lies in the low quality of Soviet microelectronic components, a point that we will discuss below. But, according to the computer scientists we interviewed, a large part of the explanation relates to the divorce between scientific research and industrial production. They argued that in the early 1960s computer science in the Soviet Union was very advanced, as shown by the applications of Soviet-made mainframes to the space program. But in 1965, when the technological race in computers started to accelerate in the rest of the world, the Soviet Government decided to start mass production of computers, and shifted the budgeting and technological decisions to the industrial and military ministers. Lacking substantial scientific basis, and not wanting to surrender to the technological control of the Academy of Sciences (a bureaucracy with its own agenda), the industrial ministries and the defense-related firms, under pressure, decided to copy American computers and chips using reverse engineering. They copied IBM and DEC mainframes and INTEL chips.

From that point, the Soviet computer industry became dependent on its ability to copy American computers already on the market. It thus came progressively to lag behind American technology; it never mastered the basic research behind each new computer generation, and – driven by pressures of government, particularly military, demand – it had no incentive for cost-cutting and manufacturing quality. Soviet researchers believe that had fundamental research along a different line of algorithms been pursued, it could by now have produced a different brand of computers, at least comparable in quality with Western technology.

But, once technological dependency on Western computers was built into the system, during the 1970s, computer penetration in the Soviet Union fell dramatically behind the West; the combined impact of lack of hard currency to import computers and parts, and COCOM controls on high-technology trade with the Soviet Union, severely hampered the ability of the Soviet economy to enter the information age. Furthermore, the concentration of most of the advanced imported equipment in the secluded military sphere made it increasingly difficult to diffuse advanced computer technology in industry and more widely in the Soviet economy and society.

A similar tale could be told for microelectronics. Here, again, the isolation from technological developments in other countries, and the lack of interaction between scientific research and industrial applications in the Soviet Union, considerably hampered Soviet progress. The extent of the failure can be judged by the fact that, as far as we can assess, there was no design capability for sub-micron integrated circuits as late as 1990, either in Siberia or in the advanced electronics center of Szelenagrad, near Moscow.

In Akademgorodok a key institute for industrial applications of electronics is the Institute of Automation and Electrometrology, established in 1957, with the intention of supporting the Siberian machine-tool industry. In fact, as already shown, this link-up never took place; the Institute emphasized other lines of research, mainly in physics – following more traditional academic fields established in the Academy of Sciences, particularly theoretical non-linear physics – and in laser research. The Institute did develop some potential technological applications, such as high-power ionic lasers (500–600 watt light power, 30 times more powerful than current industrial lasers), and crystal lasers based on ultra-shock pulses. It also branched out into research on optical memory mechanisms and photosensitivity; and it built measurement devices for use in precision machinery, and remote sensing devices for robotics applications. It invented and built the most accurate gravimeter in the world, currently used in several countries. And it developed 3-D images generating computer graphics systems, to be used in training systems for aircraft and space programs. Yet, in spite of this impressive list of potential direct applications for electronically-based tools and systems, the Institute rarely received requests from industrial firms for its technology.

The reason, according to the researchers, was that industrial firms were not interested in state-of-the-art technology; their production plans were adjusted to the machinery they already had installed, and any change in the production system would mean failure to meet the production quotas assigned to them. Therefore, technological change could happen only through the impulse of the corresponding Gosplan unit, which would have to order the introduction of the new machines at the same time that it determined a new production quota.

But Gosplan's own calculations could not rely on potential machinery that might result from leading-edge research in the academic institutes. Instead, Gosplan relied on off-the-shelf technology, available in the international market. This is why it tended to copy or buy Western machinery to modernize Soviet factories, ironically undercutting any chance for industrial application of Soviet academic research. As a result, electronic machinery and microelectronics progressively became extremely backward in the Soviet Union, and this state of affairs has come to affect the entire system of information technology devices, with a huge negative impact both on the productivity of industry and on the quality of scientific research.

It is hardly too much to say that this failure to apply innovation proved to be the Achilles heel of the entire Soviet economy, guaranteeing its eventual downfall. In any event, the ultimate consequence of the Akademgorodok experiment was a pleasant, peaceful, remote community of aging, somewhat disillusioned scientists, routinely working side by side, without effective communication, in stand-alone bureaucratized research institutes, dependent upon the administrative rule of the Siberian Branch of the Academy of Sciences, and effectively secluded from backward and forward linkages with the industrial world.

However, history has its surprises. And when *perestroika* arrived in Akademgorodok, though it brought shortages of food and medicine, it also revived great projects and grand dreams. Researchers and community leaders alike started to talk about technopoles and Silicon Valleys on the banks of the Ob. It might seem that, after all, it is not easy for the seeds of innovation to be blown away, even by the harsh Siberian wind.

From science city to technopole?: entrepreneurial Akademgorodok

The transition to a market economy in the Soviet Union fundamentally shook up the quiet world of Akademgorodok. Institutes, knowing they possessed valuable scientific knowledge, sensed the possibility of marketing their products and their processes, using the proceeds to increase their institutional budgets, to win administrative autonomy, and, last but not least, to substantially increase their scientists' salaries. And these scientists saw the opportunity to improve their living standards and at last to obtain some personal reward for the work they had done over the years, with so much effort and so little recognition. Cooperatives, self-accounting units, and sheer moonlighting enterprises started to proliferate in Akademgorodok, often linking up with similar entrepreneurial ventures in Novosibirsk, Kemerovo, Krasnoyarsk, and other Siberian cities; start-up cooperatives and entrepreneurial individuals became active hustlers in seeking foreign partners to finance projects to exploit their research (and sometimes that of others) in the Academy's institutes.

It seemed as if, in a paradoxical twist of history, Akademgorodok could at last effectively become a scientific complex able to spur new economic growth; and thus, projects for technopoles, some of them pragmatic, others visionary, started to appear in the discourse of some academic innovators. The response of the Academy leadership became increasingly confused: it oscillated between trying to arrest the disintegration of academic research, and taking the lead in the transition to the new relationship between science and economic development. In between, the remaining serious scientists seemed to regret the lack of an effective breathing-space between the two millstones of bureaucracy and market forces.

The institutes of the Academy took advantage of the new flexibility, and the loss of control by the central administration, to increase their financial autonomy by relating directly to industrial enterprises that could use their services or buy their products. In reality, the most powerful institutes had enjoyed this freedom for some time; thus, as already noticed, the Nuclear Physics Institute produced advanced experimental machinery that it then directly sold to other scientific institutions and to industries in the Soviet Union and abroad, using the funds directly to supplement its budget. For example, dealing directly with another institute of the Soviet Academy of Sciences, and bypassing the Siberian Academy bureaucracy, it built a linear collider in Moscow. In the new circumstances, it expanded its market connections in the European regions of the Soviet Union, and in 1990 it was building two centers for synchotron radiation systems in Moscow and in Szelenagrad, again circumventing the Academy.

Similarly, the Institute of Tomography, and the Institute of Automation and Electrometrology, sold their equipment to other scientific and industrial units. Increasingly, institutes related directly to industrial enterprises, developing specific products to direct order, according to specifications. It is important to notice that such linkages, which began only in 1990, completely bypassed both the industrial ministries and the Academy of Sciences, and could only work because of the direct, in fact illegal, links between firms and researchers. But the research undertaken under such conditions did not follow any long-term plan, and certainly did not respect the guidelines of the Academy for a given institute in a given year. Yet, the work was performed in the working time and with the equipment of the institutes; a practice that created considerable problems, as the administration moved to take control back from the researchers as they ran amok, earning much more from outside consulting than from their regular salaries.

The institutes went further, not merely increasing their resources through sales, but establishing cooperative firms that would commercialize their products. This was particularly the case with the new, important Institute for Informatics Systems, established in 1990, with over 200 researchers in 10 separate laboratories. This institute immediately set up a cooperative firm, INTECH, to commercialize its research, paying taxes on the profits earned, giving back some of the earnings to the Institute budget, and retaining another portion for the researchers involved in each specific transaction. Its market comes mainly from industrial enterprises in the European regions of the Soviet Union, to whom they sell computer services. In 1990, also, it set up two joint ventures with foreign companies: POLSI (Poland–Siberia); and DIALOG, a Soviet–American venture, to commercialize its software products abroad, as well as foreign programs in the Soviet Union. In addition to computer programs, INTECH also sells computer hardware that the Institute produces for its own research. Overall, INTECH is allowed to keep only about 7 percent of its total earnings. But this is enough to provide budgetary flexibility as well as substantial incentives to the researchers. As a result, the Institute's director confessed to us that research orientations, particularly among young scientists, had fundamentally shifted away from basic research to commercial applications.

But the majority of commercial spin-offs in Akademgorodok in 1990 were uncontrolled by the institutes. They were the work of scientists themselves, most of them still in their thirties, who were forming cooperatives and self-accounting groups to market the products they were developing in their institutes, as well as consultancy services, sometimes only marginally related to their central scientific activity. There were reportedly hundreds of these start-up firms in Akademgorodok and Novosibirsk. One such was NABLA, formed by a group of researchers from the Institutes of Chemistry, Mechanics, and Geology. Their main activity was to sell scientific instruments, designed and produced by themselves, new materials, and computer consulting to industrial enterprises and to academic institutes. This case illustrates that under commercial conditions, and bypassing the Academy's controls, researchers were relating to the activity of other institutes. They made major discoveries in plasma chemical processing, and invented a new plastic material for using in catheters, as well as synthetic materials for metal printing.

Overall, they diversified their production as much as possible, to start a process of long-term capital accumulation and reinvestment. First, in 1988, they formed a cooperative, named "Analysis." Then, after the Government stepped up controls on cooperatives, they moved to become a self-accounting unit within a State Cooperative. This allowed them to hire workers, either permanently or temporarily, on contracts specific to a project. Their salaries in the firm were about 900 rubles a month, compared to 400 rubles a month as heads of laboratories in the Academy.

NABLA researchers acknowledged that they used time and equipment from the Academy, and that the bureaucracy was unhappy about their activity. But they also said that their institutes could do nothing to control their activity, given that, by the prevailing academic standards, they could achieve above-average scientific productivity merely by doing some research in their spare time. They justified their move, in spite of their deep interest in basic science, because of their disappointment with the institutionalization of science in Akademgorodok. They accused the Academy of feudal working relationships, lack of scientific commitment, and lack of stimulus and communication. They claimed that, left to the usual routines of life in Akademgorodok, they would go insane.

Thus, the researchers reacted by trying to become capitalists. Their concern was the lack of dynamic market relations in the Soviet Union. So their obsession was to establish joint ventures with foreign companies, to commercialize abroad their scientific discoveries: they were convinced of their ability to innovate and of the quality of their products, and were desperately looking for foreign partners.

In the meantime, they had decided that in prevailing Soviet conditions it was too risky to invest all their earnings in advanced-technology products. So, with the earnings from their first year of activity, they made two new investments: a ship for commercial transportation on the Ob river; and a small sausage factory. In their own words, in a personal interview, they said, "If things continue this way, there will be not much use for technology in Russia, but people will always need sausages. And if the situation gets even worse, we can always use our boat to escape . . . (sic)." It would seem, in fact, that NABLA's young entrepreneurs are quite representative of the new breed of would-be scientific capitalists, who are spinning-off from the research institutes at an increasing pace in post-communist Russia.

This flurry of entrepreneurial activity led some researchers to suggest that the time was ripe to propose the transformation of Akademgorodok into a technopole, understanding by that term the establishment of a network of linkages between scientific research and the marketplace through a series of intermediaries, made up of information services and consulting firms. Interestingly enough, two such proposals have emerged from the Institute of Economics and Industrial Organization, the leading Institute that from the 1960s onward, under Khrushchev's original impulse, designed the original, ambitious regional development plans for Siberia. Thus, the cooperative firm SIBCONSULT, a spin-off supported by the directorship of the Institute, proposed the creation of an *ad hoc* institution to commercialize scientific discoveries, developing their commercial applications. The institutions will reinvest the earnings in developing new scientific infrastructures, and in supporting commercially-oriented scientific projects.

Examples would be the Institute of Chemical Rinetics, specializing in proton emission for treating cancer; the Tomography Center; and SIBCONSULT itself, commercializing its specialized knowledge in regional development and management training in the service of industrial companies and coal-mining enterprises that seek foreign partners and foreign markets. The Institute proposed the development of high-technology production in Siberia, exploiting the pool of high-quality Siberian scientists and engineers earning very low wages by international standards. Their development plan proposed three stages of growth for the new Siberian economy:

1 clarify the legal aspects of property and market relations;
2 carry through a managerial reform of the Siberian enterprises;
3 convert the military-related firms, the most technologically advanced, by linking them up with the Academy Research Institutes to form the backbone of an advanced industrial economy oriented toward export markets.

Under such conditions, Akademgorodok would become the scientific core of the new development process, once the Soviet economy establishes its linkages with the world economy.

Another group of researchers in the same Institute of Economics and

Industrial Organization were thinking further in 1990. They proposed the creation of a real technopole, a new town, a few kilometers from Akademgorodok. These scientists believed that the structure of the Academy was so bureaucratic, so conservative, that no substantial change could happen under its auspices. So they presented a project – approved in Moscow by the State Committee on Science and Technology, but predictably rejected by the Siberian Branch of the Academy – to start a techno-village, made up from spin-off companies from the Academy's institutes, as well as from some new, more dynamic research institutes. They claimed that about 500 such firms existed already in Akademgorodok, but that they were not visible, nor were they interconnected, because they still lived in the shadow of the Academy. In their vision, Techno-Village would bring together these scientific start-up companies with foreign partners, so that exchange of information would take place, along with a new emphasis on export–import markets. Thus, the foundation stone of Techno-Village should be an international hotel, capable of housing the first foreign pioneers arriving to take part in the development of the new Siberia. They claimed to need only $1.5 million to start the project. And, in 1990, they were looking for partners and for the support of international foundations.

Thus, in 1990, powerful new forces in Akademgorodok were actually reinventing the original 1960s vision of a communal development utopia. The difference is that this time around, horizontal entrepreneurial networks, not vertical State institutions, would form at the core of the dream. Yet it is significant that Siberia continues to project itself into the future, as a growth machine synthesizing an aspiration toward a communal, natural life (the new town set amidst nature) with the power of science as the engine of development (the research institutes). Statism and capitalism are both subsumed in this peculiarly Siberian vision of originating the world of the future in the immense *taiga* secluded by the perennial winter.

The real lesson of Akademgorodok

The Akademgorodok experiment failed as a regional development project, as an instrument of technological modernization, and as an attempt to create a scientific complex. The quality of the research in its institutes was very high, simply because of the quality of the scientists who went to work there. But little added value resulted from the spatial proximity between the institutes, as they rarely related to each other.

Thus, scientific-technological complexes only become sources of innovation and factors of growth on the basis of linkages and interaction between their different components, and between the complexes and a broader productive structure. The organizational logic of the Soviet economy, embodied in the administrative structure of Gosplan, the ministries, and the Academy of Sciences, forbade the formation of a true complex of technologically-led production. The structure and logic of the Academy itself gave little incentive to forge connections between the scientific units.

As the pioneer enthusiasm of the Khruschevian dreams dissipated, and the grayness and repressive atmosphere of the Brezhnev regime settled over the complex, Akademgorodok became a trap for those who could not leave, feeding itself on the intellectual blood of a new generation of Siberian scientists and indulging in its own myth as a pioneering innovation, increasingly closed to the outside world. The quality of life deteriorated, as housing shortages began to bite, promised amenities did not materialize, social life atrophied, and nature revealed its true inhospitality, symbolized by the long and harsh Siberian winter.

And yet, as soon as *perestroika* opened up the possibility again to link science and economic life, there came a revival of the dream: technology and industry, working within a primitive quasi-market economy, and using a wealth of entrepreneurial initiatives and scientific ingenuity, asserted the potential for growth and innovation that had been so long stifled by bureaucracy and reinforced by apathy. It may now be too late for Akademgorodok to become a truly innovative technological center. But if Siberia ever awakes to its fantastic developmental potential, under the new conditions, it seems likely that Akademgorodok and its spin-offs will form an essential ingredient of the Russian version – a frigid version, perhaps – of the new informational economy. Spatial concentration of scientific resources is not a sufficient condition to spur technological innovation and economic development. But it is indeed a necessary one.

SCIENCE TOWNS AND HIGH-TECHNOLOGY INDUSTRIALIZATION: TAEDOK, SOUTH KOREA[3]

South Korea is the country that, in the last quarter-century, has most rapidly advanced from industrial and technological backwardness to become a major power in electronics manufacturing and in some areas of research and design in information technology.[4] It has dramatically upgraded the technological foundations of its manufacturing industries and the quality of its technical and engineering workforce. At the roots of this success story lies a developmental state, able and willing to increase the competitiveness of Korean companies in the world economy by providing necessary support in the form of industrial and communications infrastructure, labor training, credit, trade policies, and science and technology.[5] Total national investment in science and technology, as a proportion of GNP, grew from 0.26 percent in 1965 to 0.86 percent in 1980, and then jumped to 1.77 percent in 1985. Furthermore, the proportion of government to private funding of such science and technology investment, 9:1 in 1965, fell to 3:7 in 1985, showing the increasingly strategic role of technology in the building of industrial competitiveness.

The first round of technology policy was undertaken by the Ministry of Science and Technology (MOST), created in 1967 with the responsibility to coordinate the technology programs of different ministries. But such coordination was not an easy task: other ministries considered it an encroachment on their own turfs. So a National Council for Science and Technology, under the chairmanship of the Prime Minister, was established in 1973, with MOST assuming the secretariat. In 1982, this Council developed into a new and more powerful institution: the National Technology Promotion Conference and the Technology Promotion Council, chaired by the President of the Republic himself. The Conference, comprising about 200 members, includes high-level officials, industrialists, and scientists; the Council, formed by vice-ministers and appointed members, deals with the implementation of the policy. MOST again provides the secretariat of both institutions, and the Science and Technology Minister presides over them on behalf of the President. Thus, the necessary political muscle was given to technology policy in the critical period of Korean development.

The Government exercised strategic guidance over both technological and economic development at national level, particularly through the Economic Planning Board. The shift from light industries to heavy industry, then to high-technology manufacturing, was planned by the government and executed by large Korean conglomerates (the *Chaebol*) whose formation and orientation had been decisively shaped by the State.[6] However, social and territorial planning fell far behind in government priorities. Although a regional development policy existed on paper, aiming at decentralization of economic activity and regional balance, the instruments for its enforcement were all too weak.[7] Particularly, high-technology manufacturing became increasingly concentrated in the Capital Region (the Seoul–Incheon–Kyonggi area), which, in 1987, accounted for 43.6 percent of national manufacturing output, and for 62.8 percent of high-technology manufacturing output (Figure 4.3). The concentration of R&D functions is even higher.[8]

The Government was aware of the need for regional decentralization, particularly in the context of acute regional rivalry in South Korean politics. Thus, given the symbolic value of R&D facilities, in the early 1970s President Park himself took the decision to propose a new science town to be built in central South Korea, on an undeveloped site in the heart of the country: Taedok Science Town was born.

Vision to reality

Taedok was entirely conceived, built, and for many years managed, by the Central Government and its subordinate local agencies, through the Ministry of Science and Technology. It is located 100 miles (160 kilometers) south of Seoul, near Taejon in the middle of the Korean countryside. It can

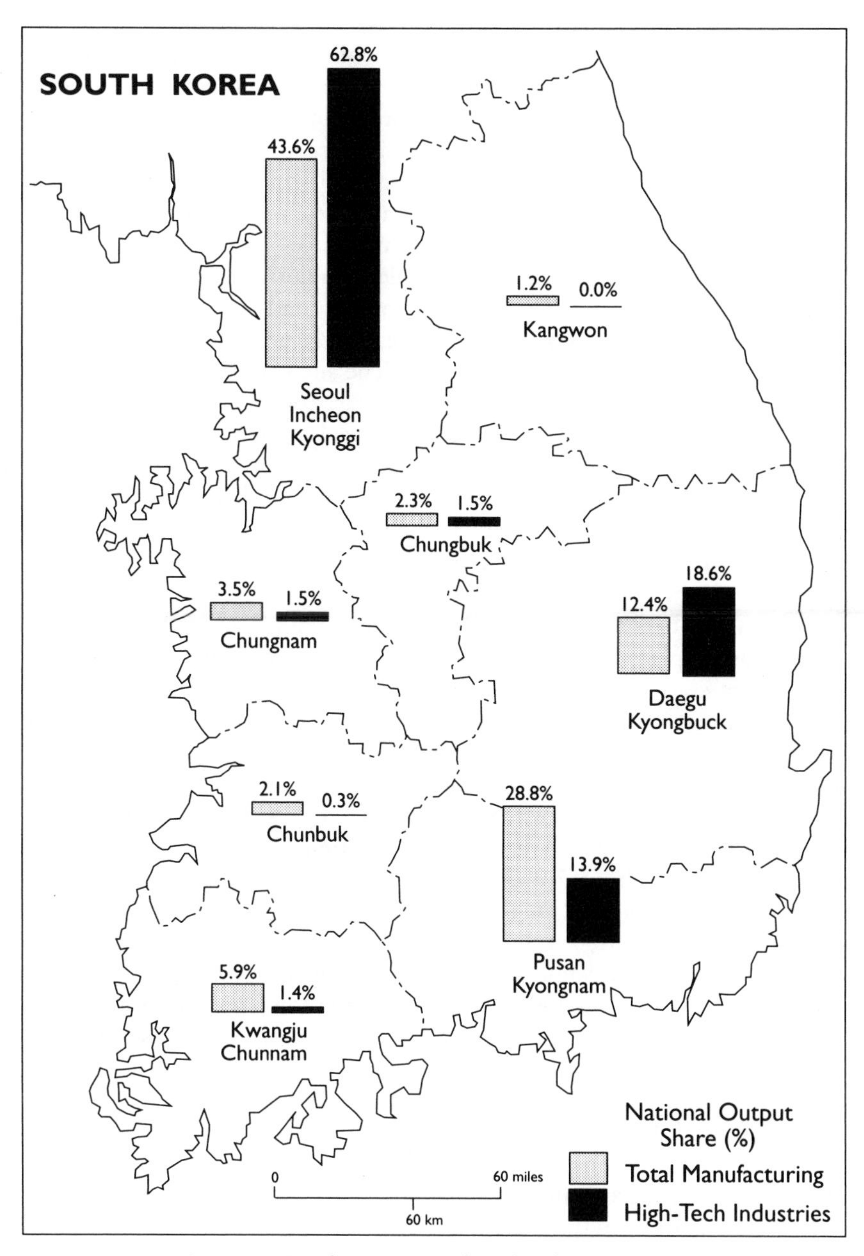

Figure 4.3 South Korea: High-technology industry
Source: Anon, 1990

be reached from Seoul, in about two hours, by freeway; there is also a "rapid train" (certainly not a high-speed train) from Seoul to Taejon. The Science Town also has an acceptable level of telecommunications infrastructure, connecting it with the automatic international network. It covers 1,134 acres or nearly two square miles (2,800 hectares), of which 46 percent is for research institutes and university facilities, 7 percent for residential purposes and community services; the rest (47 percent) is open space. The town is a distinctive spatial unit, entirely segregated from the city of Taejon (1 million people), 18 miles (30 kilometers) away.

In 1988 there were only two possible entry points, both guarded by the army at road checkpoints. The construction is of rather good quality, both for the scientific institutions and for the residential areas, which are laid out in the midst of forests and garden spaces. The site is green and hilly, and is quite pleasant without being beautiful. There are some high-class areas, with single-family dwellings for the directors and senior scientists of the research institutes. The quality of housing and urban services is incomparably higher than that of Seoul. But the town is built at very low densities and appears entirely lacking in urban life, at least when we visited it in November 1988. The main social activity takes place in the city of Youngson, five kilometers (three miles) away, which is a thermal station with some hotels, and a few restaurants and bars that provide the only meeting places in the area.

The master plan for Taedok Science Town was approved in 1973; construction started in 1974, and the first institutes moved to the town in 1978, with the relocation of four government research institutes from Seoul. The year 1985 saw completion of the west section of the town, the first stage of the project. In 1986 building began on the east section, the second stage. Completion of the whole town, as initially planned, was scheduled for 1992. Thus, it will have taken about 20 years to complete the project. In 1988 Taedok contained about 7,000 scientists, technicians and research-related personnel. On completion the town is planned to have some 10,000 science-related personnel, with a resident population of about 50,000 people in 10,000 housing units. Some of the labor force, particularly manual and service workers, are expected to live outside the town and commute. Thus Taedok is expected to be a highly selective residential area, reserved only for scientists and managers, who will enjoy excellent schools for their children and better services than they had in Seoul.

In 1988 Taedok housed no less than 14 research institutes and three universities. Of the 14 institutes, nine were government institutes, two were institutes with government participation, and only three private. About 1,000 Ph.D.s worked in these institutes in 1988. The three universities are all quite specialized: CHJCCS, specializing in computer sciences; CHNU, a science-oriented university; and the Korean Institute of Technology (KIT), an elite institution focusing on undergraduate training in engineering for a nationally-selected group of young students.

In addition, the Government foresees the relocation to Taedok of Seoul's Korean Advanced Institute of Science and Technology (KAIST), the most important science and technology institution in South Korea. However, given the strong resistance that we observed at KAIST when we interviewed its members in Seoul, and given the logistical problems involved in relocating the entire institution – housing for at least 10,000 graduate students, for instance – it seems most likely that only certain programs and functions of KAIST will be relocated to Taedok.

The town is served by a Multiservice Center, of some 40,900 square feet (3,800 square meters), including 30 service units, such as a hospital, a cafeteria, post office, bank, pharmacy, shopping center, bookstore, and restaurants. There is an indoor swimming pool, and a golf course was in the planning stage. The entire complex conforms to a new spatial concept, developed by the Korean Government, of "technotopia." Thus there is no manufacturing or agricultural activity in Taedok; and there is absolutely no connection between the research activities in the town and the productive activities in the area around it, particularly in Taejon, a relatively industrialized city.

The administration is rather simple. It depends entirely on the Ministry of Science and Technology, which delegates its power to the Administration Office of Taedok Science Town. This Administration Office is composed of two divisions: Development and Management. The Development Division is in charge of building the city, thus implementing the master plan. It also coordinates all the government agencies involved in construction of the city, and provides material support to the functioning of the institutes. The Management Office allocates and manages the residential units and provides the necessary services. The budget is decided in Seoul by the Economic Planning Board after consulting with the Ministry of Science and Technology. The city of Taejon is in charge of building and maintaining access roads to Taedok, as well as providing and managing water and sewage services. The Ministry of Construction is in charge of public works within the town, and electricity and telephones are of course serviced by the public companies. Thus, the key institutional features are the lack of local autonomy *vis-à-vis* the Central Government and the entirely managerial approach of the local administration. For the public service and for the social life of the residents, a public institution (also set up by the Central Government), the Korean Science and Engineering Foundation, is in charge of securing grants and running programs.

The research institutes

Taedok was thus conceived and implemented as a pure-science town entirely devoted to research institutions, supported by scientific and engineering universities. The intended linkage with private firms was to be established through the research organizations of the private sector. But down to 1988, the Government was clearly the main actor, because 11 of the 14 research institutes relocated in Taedok were government or public research institutions.

Analysis of the content and characteristics of the institutes shows that while the intention of the plan was, and is, to attract private research institutes, so that ultimately about 40 percent of the Science Town would be based on private firms, the launching of the project was only possible because the Government literally forced public research institutes to move to Taedok. Thus, if eventually Taedok becomes an R&D center, it will be the result of a government push, rather than of an entrepreneurial pull.

The research programs also demonstrate a great deal of heterogeneity. Ginsing and Tobacco Research goes hand in hand with Ocean Engineering (far away from the ocean), and with Security Printing and Minting. It seems quite evident that there has not been any scientific plan to provoke synergy among these institutes, simply generalized pressure – applied to both public and private research institutions – to go to Taedok, regardless of the substance of their research. Considering the list of possible future private research institutions, the area of information technology (paramount among the priorities of Korean R&D) seems to be underrepresented. Instead, the list includes a large number of traditional manufacturing research institutes (textiles, chemicals, fertilizers, shipbuilding, tires, etc.), probably indicating that private firms are acceding to government pressures by sending to Taedok only those research agencies whose level of technological sophistication can be divorced from the main research and manufacturing centers in the capital region.

Our own interviews with the institutes in Taedok confirm this hypothesis. They all report serious complaints from their researchers after being compelled to move to Taedok, and many of them keep their families in Seoul, commuting on weekends. The scientists' main anxiety is the quality of schools for their children, a major concern in Korean family life. The institutes have minimal contact with each other, and they all report that they could conduct their activity anywhere; they get no benefit in interaction or synergy from their location in Taedok.

As far as we can assess the situation, it would seem that Taedok constitutes no scientific milieu; that there is no scientific policy behind the planning of the research institutes moved there, and no synergy produced by the spatial contiguity of these research activities. Taedok is a pure manifestation of an all-powerful government's will to relocate those institutions over which it has direct or indirect power, and which can be relocated without major harm to their scientific output.

Location and milieux of innovation in Korea

The progress made by South Korea in science and technology has relied on a combination of major efforts by the Korean *Chaebol*, supported by the

Government, and on a program of creation of universities and research institutes, public and private, most of which were traditionally located in the capital region. No less than 464 out of 674 enterprise research institutes are in the capital region, as well as 13 out of 24 government institutions, 39 out of 40 research associations, and 47 out of the 104 universities and colleges existing in South Korea.[9] The main reason for this concentration is the availability of high-level engineers and scientists in the Seoul area, particularly for the most advanced of these scientists, many of them returnees from the United States and Europe. Seoul is clearly the only place in South Korea where they can find themselves connected to the wider world of science and technology.[10] Thus, in spite of the poor quality of life in the congested South Korean capital, the real technopolis in Korea is Seoul, and particularly the suburban belt around the city.

In this context, the Taedok project appears as a purely political decision, in the early 1970s by the then-President Park, to locate a new center of excellence in his home province (Chungnam), so as to mark his efforts at decentralization, and to provide a political counterweight to the influence of the restive regions of Chunnam (capital: Kwangju) and Kyongnam (capital: Pusan), the centers of the opposition to his regime. Neither functionally nor industrially was Taedok a viable project. Thus, in spite of the incentives offered to potential tenants, and the relatively pleasant atmosphere of the site, after 15 years the only institutes located there were those compelled to relocate from Seoul, either because they were part of government or (in the case of the three private institutes) they were established as a *quid pro quo* for government help for their parent corporations.

Two serious consequences followed. First, there was practically no scientific synergy among the institutes, and therefore no milieu of innovation formed, at least in the first two decades of Taedok's existence. Second, no linkage or feedback developed with manufacturing or applications of any kind. Thus the "learning by doing" approach to technological development, so crucial in the Korean experience, could not even be attempted. The isolation of Taedok *vis-à-vis* the industrial environment of Taejon was deliberately sought at the beginning, as a way to inject a new regional dynamic into the area. In fact, it achieved the opposite: it helped establish an artificial implant, without the capacity to promote development.

Furthermore, the city of Taejon and the Provincial Government were generally hostile to these technocrats from Seoul, who consumed resources without contributing to the area's finances, and who bypassed all local and regional powers via their direct ministerial connections in Seoul. Thus the overall result, from a 15-year perspective, appears a typical case of failure of regional decentralization policy, which arose in consequence of an arbitrary political decision taken without reference to the logic of the market, to the dynamic of technological development, or to the wishes of the engineers and scientists involved.

Yet, once Taedok was in existence, a second level of analysis was needed. Granted that it was an entirely artificial creation, the concentration in one area of thousands of scientists and engineers, with a vocation for industrial research, does create new opportunities. Thus, using just such an opportunity, in 1989 the city of Taejon planned a "High-Technology Industrial Park," to be located in an agricultural area of about 154 square miles (4 million square meters), next to Taedok. The park will focus on semiconductors, the biomedical industry, and new materials. It is expected that Taedok will provide most of the research basis for this development.[11]

The attempt to develop synergy is a central element in a comprehensive program announced by the South Korean Government in 1989: the "Technobelt" concept, to link up research and industry throughout the country, via communications and telecommunications.[12] In this program, the "epicenters" of such a "technobelt" system would be the existing science centers in Seoul and in Taedok. From them, technological diffusion centers would spread connecting networks along four major "Technobelts" that would integrate most of the country into the productive structure of the twenty-first century (Figure 4.4):

1 A West Coast Belt, from Seoul to Kwangju, specializing in energy and food production;
2 A South Coast Belt, from Kwangju to Pusan, centered on heavy industry;
3 A North–South Belt, linking Suwon, Taedok, and Pusan, organized around electronics, semiconductors, and textile industries;
4 An East–West Belt, connecting Kwangju, Taedok, and Kangnung, with new developments in medical and social services research and production.

In such a "technobelt" system, Taejon plays a major role through its central location, and Taedok becomes the research center of such an innovative milieu, this time strongly articulated to manufacturing activity.

Regardless of the feasibility of such strategic regional planning, which is typical of the current futuristic state of mind in the Asian Pacific, what seems significant is that Taedok, in spite of its isolation and lack of historical justification, may become an engine of future regional development, just because it exists. Whether as the "epicenter" of a "technobelt," or as the research backbone of a new industrial area focusing on the new manufacturing sectors within the Korean economy, Taedok seems to be on course to become linked to a highly productive segment of the world economy, and – through such a process – to become a resource for regional development. The new mining sites of science and technology may appear to be the pure result of the caprice of a Prince, or of the visionary dream of a technocracy.

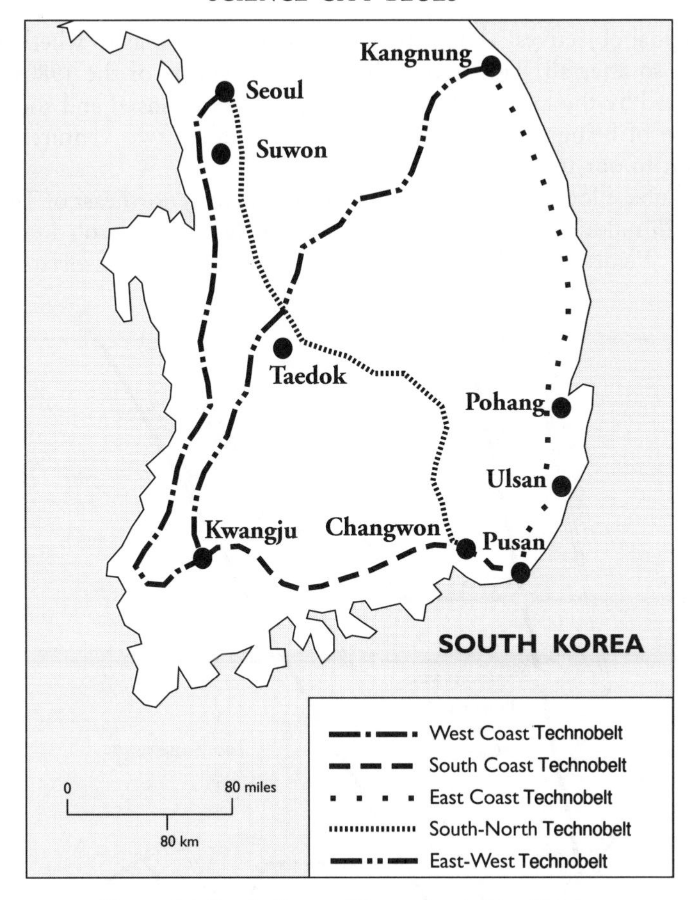

Figure 4.4 South Korea: The technobelt policy
Source: Anon, 1990

Yet, as soon as thousands of scientists, engineers, and students concentrate in any such site, and as long as they continue with their lives and their research, the new productive forces of our era could develop there at any time. True, a science center might have been created more cheaply, more efficiently, more speedily, and with less human disruption. But, at the end, Taedok will exist, as a science town linked to an industrial area. In forming an innovative milieu and then shaping its outcome, the genesis thus proves less important than the subsequent logic of development.

SCIENCE CASTLE: TSUKUBA, JAPAN

Japan has made two major attempts to build a science city. The first, Tsukuba, conceived in the 1960s, was in origin totally a government-promoted scheme, and seems to have had a great deal of difficulty in meeting its original objectives – but with a remaining question as to whether it may not do so after all. The second, Kansai, is a scheme of the 1980s; though promoted by the local prefectures, it is privately financed and shows every promise of being a success. So the stories of these two ventures are very relevant to our trail of inquiry.

Tsukuba is located about 40 miles (60 kilometers) northeast of Tokyo and about 25 miles (40 kilometers) northwest of Narita, Tokyo's international airport (Figure 4.5). Though the outside world often associates it with

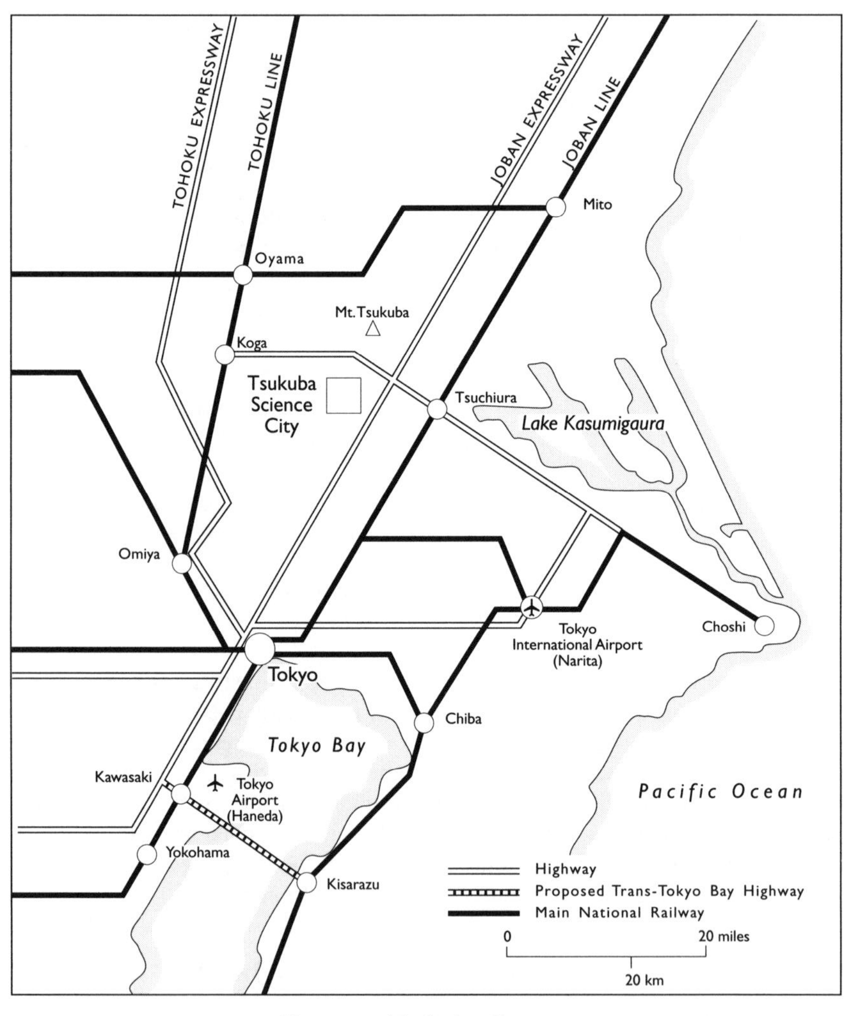

Figure 4.5 Tsukuba: location
Source: Nishioka, H. (1985) "High Technology Industry: Location, Regional Development and Internal Trade Frictions," *Aoyama Keizai Ronshu*, 36

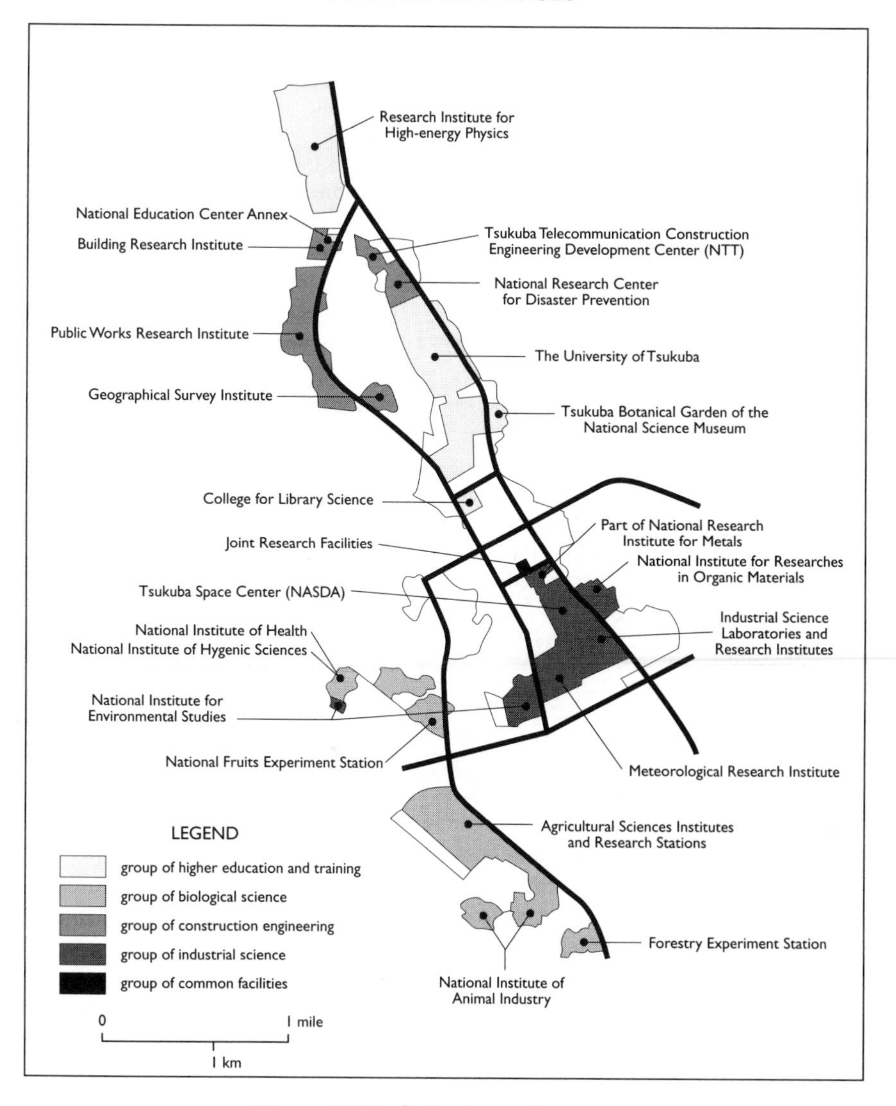

Figure 4.6 Tsukuba: internal structure
Source: Bloom and Asano, 1981

Japan's technopolis program (Chapter 9), it was built much earlier, for different purposes – though it did later serve as a model for several technopolises. It is a national research center, funded totally by central government; its laboratories, engaged on basic research, are government laboratories.[13]

It is a big development, covering 112 square miles (28,559 hectares), about half the size of the 23 wards of Tokyo. A core area of some 10.4 square

Table 4.1 Tsukuba: a brief history

Year	Event
1961	The Cabinet decided to relocate government offices collectively outside Tokyo in order to prevent the area's overpopulation, and to examine concrete measures to achieve it
1962	The Council for Science and Technology, an advisory body to the Prime Minister, proposed collective relocation of national research institutes for their revitalization The Cabinet approved a plan to develop about 4,000 hectares in the Tsukuba area as a site of relocation, charging the Japan Housing Corporation with acquisition and preparation of land
1964	A blueprint for the Academic New Town was completed by the National Region Development Commission The Cabinet approved of a plan to start the construction from 1965 and to complete it within about 10 years The Cabinet decided to establish Headquarters for the Development of the Academic New Town in the Prime Minister's Office that was headed by the chairman of the National Region Development Commission
1967	The Cabinet approved of the basic policies of the development of the New Town, and the number of national institutes to be transferred to be 36
1968	Construction work of research/educational facilities started with a large-scale earthquake simulator at the National Research Center for Disaster Prevention of the Science and Technology Agency
1969	The Cabinet decided to complete transfer of 11 national institutes by 1972
1970	The Tsukuba Academic New Town Development Law was enacted
1971	The Headquarters for the Development of the Academic New Town finalized fundamental principles of the development project and the outline of the construction program of public and service facilities
1972	The National Research Institute in Inorganic Materials began operation, the first institute to start The Cabinet decided to increase the number of national institutes to be relocated or newly established to 43
1973	Tsukuba New Town Development Company was established with a joint investment of public and private sectors. The company was entrusted with the management of parks and roads constructed by local public bodies as well as the construction and management of shopping centers The University of Tsukuba was opened
1974	The Land Agency was established and its Metropolitan Areas Development Bureau took charge of overall coordination and adjustment of the development project
1975	The Cabinet decided to change the final date for the completion of the transfer of all research/educational institutes from 1975 to the end-fiscal year 1979 The Headquarters for the Development of the Academic New Town provided the "Main Principle of the Special Measure for Financial Burdens of the Towns and Villages in the Tsukuba Academic New Town" in order to reduce financial burdens of the six towns and villages during the period of 10 years starting from the fiscal year 1976

Table 4.1 Continued

1977	The Tsukuba Academic New Town Council, composed of Ibaraki Prefecture, towns and villages concerned, research and educational institutes in the area, and Japan Housing Corporation, was established
1980	All of the 43 national institutes originally planned were in operation in Tsukuba
1985	Joban Highway was directly connected to Tokyo An international exposition, Tsukuba Expo.'85, was held, attracting more than 20 million people
1987	Five townships (Oho, Sakura, Toyosato, Tsukuba, and Yatabe) were merged into Tsukuba City.

Source: Fukyu Kouhou Senmon Iinkai, 1989

miles (2,696 hectares), slightly under one-tenth of the entire city, is called the "Research and Education District" and is developed for research and educational institutes, commercial facilities, and housing. Around it, the remaining area constitutes the "Surrounding Suburban District," and is designed to accommodate private research facilities and future-oriented industries while preserving agricultural land and the high-quality natural environment (Figure 4.6).

Tsukuba had a population of 162,189 in 1989: 46,096 (28.4 percent) in the Research and Education District, and 116,093 (71.6 percent) in the Suburban District. So the city is well-advanced to its target population of 220,000 – 100,000 in the Research and Education District, and 120,000 in the Suburban District. No less than 10,000 of the total, according to a 1989 survey, were researchers, who were particularly strongly represented in the Research and Education District. Some 2,000 of the residents were foreign, and there were in addition some 1,800 short-stay visiting researchers, especially from other Asian countries.

Tsukuba has 48 national research and educational institutes, some relocated here, others newly established. They report to a great variety of government departments and agencies: seven to the Science and Technology Agency, one to the Environment Agency, six (including the University of Tsukuba) to the Ministry of Education, and two to the Ministry of Health and Welfare; no less than 13 report to the Ministry of Agriculture, Forestry and Fisheries, another 10 to the Ministry of International Trade and Industry, three to the Ministry of Transport, four to the Ministry of Construction, and two to the Ministry of Foreign Affairs. Thus, at the end of the 1980s, 30 percent of all national research agencies and 40 percent of their researchers were found in Tsukuba; about 50 percent of the total budget for all institutes is invested there. In addition, eight public corporations are located there.

A brief history

Tsukuba originated in the 1958 Tokyo metropolitan area development plan – a London-style regional planning exercise – as a satellite science city for

Tokyo; this was followed, in 1963, by a decision to decentralize government functions and promote research (Table 4.1). Several sites around Tokyo were investigated, including the lower slopes of Mount Fuji and some other attractive highland areas. Tsukuba won, for good reasons: it was within one hour from Tokyo; there was an existing city (Tsuchiura) in the vicinity; and the proximity of Lake Kasumigaura assured a reliable water supply.[14] The Cabinet approved the site in September 1963, one year before the Tokyo Olympics, and authorized the Japan Housing Corporation to develop the site; the Ministry of Construction was in charge of building the facilities.

There were protests from local residents and government workers. Nevertheless, land acquisition began in 1967 and the Tsukuba Science City Construction Law was passed in 1970, covering six municipalities.[15] Even then, there was a great deal of opposition to moving out of Tokyo, both from the laboratories and from the Tokyo Teachers' College.[16] The city had a bad image at first, and progress was slower than expected; but by 1980, the original target date for completion, some 80 percent of the governmental bodies with some 9,000 jobs had been already moved, with some 125,000 people in Tsukuba;[17] since then, the remainder have relocated.

Particularly slow, in the first decade, was the response of private industry. Though the plan provides three major research parks for private industries, at first there was little take-up. But a major change occurred after 1985, when the International Science and Technology Exposition was held in Tsukuba; private firms wanted to locate within a maximum one-hour travel time from Tokyo without serious delays, and so were encouraged by the infrastructure improvements – particularly, the new expressway – made for the exhibition (Table 4.2). The trend has been accelerated by the Law of Promotion of Research Exchange, approved in December 1987, which allows private enterprises to use the facilities of the national institutes, and promotes personnel exchange and joint ownership of patents between national institutes and private enterprises. More than 200 private research facilities are now established or planned here. As of January 1989, eight industrial/research parks, specifically designed for private investors, were

Table 4.2 Private research facilities in Tsukuba

	*No. of enterprises**	*area (hectares)*
Tokodai Research Park	37	41.4
Tsukuba Research Park Hanare	9	6.0
West Tsukuba Industrial Park	15	102.0
North Tsukuba Industrial Park	19	128.0
Kami–Oshima Industrial Park	19	40.0
Tsukuba Technopark Toyosato	26	69.0
Tsukuba Technopark Oho	n/a	34.0
Tsukuba Technopark Sakura	n/a	65.7

Note: * No. of enterprises includes those enterprises that are planning to move in
Source: Fukyu Kouhou Senman Iinkai, 1989

under way in the city. In the Fourth National Comprehensive Development Plan approved in 1987, Tsukuba and its neighboring city, Tsuchiura, are planned to act as a core of office-based activities as a part of the policy to encourage decentralization of Tokyo's functions.

This private investment reflects the fact that in the most recent period, Tsukuba has profited from major improvements in accessibility. The Joban Railway has two stations in the neighboring city of Tsuchiura, and connects with Tokyo's Ueno Station in approximately one hour and ten minutes; it is expected to be extended into the center of Tsukuba during the 1990s. The Joban Expressway is directly linked to Tokyo's Rapid Transit Expressway, giving direct access to and from central Tokyo by car or express bus; Narita airport is about one hour distant by car.

The city also benefits from investments in what the Japanese call "soft infrastructure": information and communications technology (ICT). A television cable system, run by the Tsukuba Science City Community Cable System Consortium (ACCS), serves about 25,000 households in Tsukuba. This allows households not only to receive up to 32 channels, including satellite broadcasting as well as community-based information and news; in the future it will also give access to more advanced services featuring the system's two-way communications capacities. A local area network, connecting personal computers in each facility with a mainframe in the Tsukuba Center for Institutes, is planned and at the experimental stage.

Evaluating Tsukuba

Tsukuba represents a huge public expenditure; its development cost, down to March 1990, has been ¥1.6 trillion ($1.067 billion) The question is whether the cost has been justified. On this, the only fair answer is that the verdict is not yet in.

There is no denying that, particularly in the early days, Tsukuba's growth was slow and somewhat painful. Some of the problems were transitional, though long-lasting. During the construction phase, the Japan Housing Corporation met much opposition from landowners. This forced the Government to modify the area of land acquisition, with the result that the city extended over six different municipalities, producing problems in service coordination such as an undesirable division of school districts, and inefficient management of infrastructure.[18] This however was largely remedied by the merger of five of these municipalities into one in 1987, allowing Tsukuba to develop more effectively the infrastructure needed for technology development: additional industrial parks, office space, and recreation facilities.[19] What is emerging here, we were told in one of our interviews, is a real city; land prices are rising on a speculative wave.

But national research institutes in Tsukuba still experience some problems. One is the extreme vertical integration characteristic of Japanese government agencies, which prevents research institutes from sharing facilities, thus leading to an excessive duplication of equipment. There has been poor interaction between public research and scientific establishments on the one hand and private sector enterprises on the other, leading to lack of joint researches and spin-off firms and activities. This is exacerbated by the fact that the relocated research institutes tended to concentrate on large projects focusing on target technologies, which has created a tendency to neglect basic research and applied research relevant to local industry or small and medium firms.[20]

Even today, the city is regarded by many researchers as an isolated island, remote from normal human society. Tesuzo Kawamoto of the Tsukuba Research Consortium notes that the city lacks sufficient communication among disciplines. How research communications can be developed, with networking of researchers in various centers, is a major issue, though a number of study groups meet periodically. There is a "Tsukuba Syndrome," especially among researchers' wives. Tsukuba is a difficult place to obtain information and research support: a study shows that only 8.8 percent of essential research information came from within Tsukuba, against 23.1 percent from Tokyo. Most of the research institutes tend to institutional rigidity, which does not enhance innovation or entrepreneurship.[21]

At the National Space Agency of Japan (NASDA), one of the first institutes to move here, we were told that Tsukuba is only now starting to look like a real city; at first they felt as if they were "out in the sticks." Things changed considerably at about the time of the 1985 Exposition, and they do not feel isolated any more. At first they thought they had plenty of space, but now they find they must enlarge and rearrange. At first the Government bought a great deal of land and gave it to the Land Restructuring Agency, which sold it to the institutes; but now the only available land is in the Expo area, which will go to the private sector. So they will need to rebuild *in situ*.

Similarly, at the Institute of Building Research, we got further confirmation that problems of communication remained. Tsukuba is good for international relations, we were told, but poor for relations with Tokyo: the journey took 4 hours each day, and they sometimes needed to travel twice a week. The environment was good but relations with other institutes were weak.

On the positive side, now that the dust is settled and the institutes are firmly located here, administrators see some definite advantages. One is better physical facilities. Many institutes exploited relocation as an opportunity to renovate facilities, or introduce new ones, particularly large-scale equipment, and to increase their space. In addition, computer networks were built around mainframes in the research parks of the Agency of Industrial Science and Technology, and the Ministry of Agriculture, Forestry and Fisheries; the networks have allowed more efficient data processing, data storage, and use of data bases.[22]

Another advantage is the increased exchange of information among researchers and research institutions: the number of conferences and symposia held in Tsukuba has increased. More importantly, the agglomeration of research institutes has promoted small meetings among researchers in Tsukuba, which were difficult when these institutes were scattered in Tokyo.[23] There are several concrete cases of this increased exchange and cooperation. The agglomeration of national research institutes such as the National Laboratory for High Energy Physics (KEK) and Electrotechnical Laboratory (ETL) has given rise to mutually beneficial effects upon institutes involved in related research. For example, the photon factory of KEK has been organized as an autonomous unit to allow industrial users and others outside the jurisdiction of the Ministry of Education, under which KEK functions, to construct their own beam lines. ETL uses KEK's photon factory for soft X-ray radiation lithography to develop new process technology for 3-D integrated circuits while ETL's work on superconductivity contributes to the development of high-energy generation at KEK.[24] Another example is laser research which ETL conducts in tandem with other laboratories in Tsukuba; it aims at the development of high-powered gas lasers for use in nuclear fusion and isotope separation as well as for information transmission and machine-tool fabrication. ETL also works with six manufacturers and the mechanical engineering laboratory in developing laser control and machinery technology for flexible manufacturing systems.[25]

Even now, the major question is how far Tsukuba can be made attractive to the private sector. Until the mid-1980s, the city was failing in this respect, so that the pattern of R&D activities in the city was highly unbalanced between the public and private sectors.[26] But, during the 1980s, several initiatives were taken to encourage the private sector to make better use of Tsukuba's research institutes.

First, unlike most sites chosen for the later technopolis program, an entire university was moved out of Tokyo to Tsukuba. At the university, the International Science Foundation was established by the Ministry of Education to help develop industry/academic relationships, though there remains the problem of how far high-tech industry can be attracted.[27] Second, greater interaction between the public and private sectors has been encouraged through the Research Exchange Promotion Act. And third, a special Tsukuba Research Support Center has been built by a joint public–private consortium; it aims to provide core facilities for open laboratories and research exchange, together with "incubators" for start-up firms, exhibition halls, information services, and in-house programs. The idea is that it can achieve services for its members that otherwise would be beyond the capacity of any of them.

Established in 1978, the Center first worked to develop meeting places and then the means of exchange, for instance posters announcing meetings, and a computer network for a database, electronic mail and bulletin board. Its Tsukuba Network is an on-line database to which any prefecture in Japan has access; it is connected to the university network system, and links to other worldwide networks are under study. The Center also offers logistical support for symposia and an office for contacts between laboratories and the university; it can offer specialized services such as materials analysis, worldwide patent information, and seminars featuring leading technological developments in its members' fields. In addition, it manages living quarters for foreign researchers, a service it is now expanding, as well as Japanese classes for researchers and their families.

A key example of public–private partnership is the next-generation industry program, the Ministry of International Trade and Industry's main technology development thrust of the 1980s. In it, 12 new products have been selected for development to the stage of commercialization. And in each, at least one of MITI's Agency of International Science and Technology (AIST) laboratories in Tsukuba is playing a key role in fundamental research, along with the agency's regional government industrial research institutes. Research associations have been set up by private firms participating in each of these projects specifically in order to coordinate activities with AIST. New technologies developed during the research phase are patented by the Government but made available first to those firms involved in these projects.[28]

But if the private sector is now locating in Tsukuba, it is strictly the research arm. At the National Space Agency, which contracts all manufacturing to the private sector, we were told that these manufacturing firms remained firmly locked into traditional locations: around Yokohama/Kanagawa for satellites and Nagoya for rockets. Despite very close interaction in the design stage, there was no evidence that this was pulling the manufacturers to Tsukuba. In fact the main operations for the Agency are still in Tokyo: the Tsukuba center is restricted to tracking, control and some R&D. Linkages with the Ministry are still critical, and they are resistant to moving, we were told.

A key question about Tsukuba is how far it has developed synergy between the public research institutes, the university, and the private research laboratories, so as to generate industrial spin-offs from basic research. We asked officials of the Agency of Industrial Science and Technology, a MITI institute, about this question. There was no easy answer, we were told: their job was to do research within major national plans, both basic and applied, but rather more of the latter. At the start of the Agency's history, there was a bias toward basic research, but then they perceived a lack of connection with industry. Later still, in the mid-1980s, both government and industry put a high priority on basic research again. The latest basic research was on the human frontier. AIST benefits from the "National Projects," mainly for applied work.

The Agency does joint research with companies, or contract research. If a company commissioned research it would own the results, but otherwise the results will be generally available: AIST patents and licenses it. The Japan Industrial Technology Agency takes care of patent work. AIST also interchanges with the universities, trying to open commercial channels for them. It has no special relationships with Tsukuba University; it just happens to be close by. In time, maybe the ministries would agree to a special relationship; meanwhile, the ties are personal. There is no simple relationship between the Institute and the university; each does its own research, and a natural boundary develops. They have 5–6 joint activities with the university each year; AIST has a center for this activity, but does not organize it. In addition, there is quite a lot of movement of AIST researchers out to industry or university.

We got a rather similar picture from the Electrotechnical Laboratory, one of the national institutes. It came here in 1973 and found it difficult to operate at first: it was hard to get personal communication or to find student help, because it was then perceived as being so far from Tokyo, and there were too few universities to provide research assistants. But recently, officials told us, they had developed many links both to private companies and to universities, and had attracted researchers through joint research; so they do not feel so isolated. They have more than 100 joint research projects with private companies. Every year, staff numbers are systematically cut by government, and the private sector is supposed to compensate; they had 800 staff 10 years ago, now only 600. They do some joint research with Tsukuba University. In joint research with industry, according to the Patent Law, there is a 3–4 patent restriction to one company; over this limit, any company can license the results, the company and the laboratory sharing the proceeds on the basis of company 80 percent, laboratory 20 percent. But not many people from this laboratory transfer to private companies, or the university, even after 20 years; and those who do leave go to the big companies, for USA-style "start-up" breakaways are almost unknown here.

At the Institute of Building Research, we were told that they had relationships with the university through particular institutes; one of their ex-researchers was now a professor there, but such moves occurred only "once in a while." They had few relationships with private institutes, because they had a regulatory function with regard to the private sector, which made it awkward. Like other institutes they were shrinking their staff in accordance with government policy, but – unlike the MITI institutes – private companies showed little interest in making up the losses; the problem was that they did not do high-tech research. We were also intrigued to hear that both this institute, and the next-door Public Works Research Institute, were doing quite separate research on structural failure, though the latter institute had five times the research budget, representing a traditional gap between structural and civil engineers.

Overall, then, the verdict on Tsukuba is not yet in. For a long time, it looked unfavorable. But it does appear that during the 1980s, the city's profound psychological isolation from Tokyo was at last beginning to be broken down. This, as much as anything, was because the ever-expanding boundary of the Tokyo metropolis was at last beginning to approach close to Tsukuba, threatening to turn it into yet another satellite city. But, returning into the capital along the gridlocked expressway, we had a sense of the remaining difficulties of communication. The other, and perhaps more important, fact was that there was increasing anecdotal evidence, both from the literature and from our interviews, that other barriers were breaking down too: those that separated the public institutes, the university and the private companies. The private sector was at last beginning to invest in the city on some scale; it was fully involved with the institutes in a number of ways on a number of projects, among them very important MITI national projects. One could begin to capture, in these examples, the sense that a MITI grand design was taking shape.

That underlines a point, which we shall also find to be true of the much bigger technopolis program. Japanese time scales are simply not those of other nations. Japan in effect applies an extraordinarily low discount rate to large public projects, whereby they are not necessarily expected to show an effective return for 10, 20, even 30 years. To judge them after only a decade, which is a natural reaction for most Western observers, may therefore be a profound mistake. It may indeed be the basic mistake that the West makes in competing with Japan. To this point, we shall have to return in the last section of our study.

FROM SCIENCE CITIES TO SCIENCE FLOWS: KANSAI SCIENCE CITY, JAPAN

Kansai Science City, named for the highly-urbanized region in the heart of Honshu, Japan's main island, is a huge development that has begun to take shape only in the 1980s. Including five cities and three towns in three prefectures – Kyoto, Osaka, and Nara – it covers an area of some 58 square miles (15,000 hectares) and has an eventual projected population, in the year 2000, of 380,000 (Figure 4.7). Its core, the so-called Cultural and Scientific Research District, includes a Science Research Facility, Culture and Science Research Exchange Center, public welfare facilities, residential facilities, and others; these are located in 12 distinct zones (13 square miles, or 3,300 hectares, with a target population of 180,000 people) – six of them in Kyoto, three in Osaka and four in Nara. The center of gravity of these zones is approximately 18 miles (30 kilometers) south of Japan's ancient capital of Kyoto, six miles (10 kilometers) north of the even older capital, Nara, and 18 miles (30 kilometers) north-east of Japan's second city of Osaka.

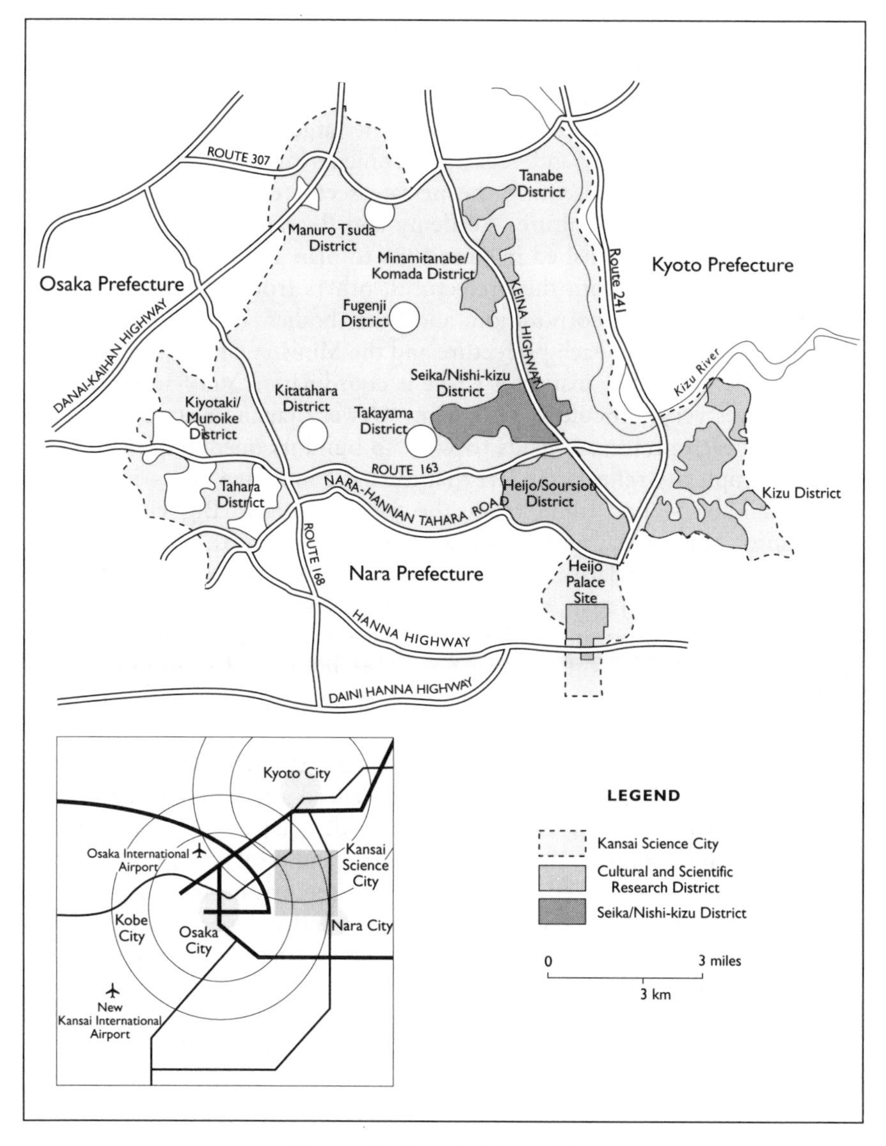

Figure 4.7 Kansai Science City
Source: Keihanna Interaction Plaza Incorporated

Only two of these zones had been completed by 1990, though many more were part-built; most will be completed between 1990 and 2000, and some are still under study. So, departing from previous conventional models – including Tsukuba and most of the technopolis developments – Kansai Science City abandoned the single-point concentrated approach in favor of a linked, multinuclear development, in which the 12 areas will be integrated via traffic and information networks into an organic whole. This is a flexible, step-by-step system which exploits the available infrastructure.

Unlike any previous Japanese model, Kansai was built from the start upon public–private partnership. As a result, its organization is complex: it involves the prefectures and local governments on one side, the Kansai Research Institute representing the private sector on the other, with the Association of Kansai Culture, Academy and Research City in the center. The Institute was established in June 1986 to plan and develop the city. It has 20 members, 10 from the prefectures, others from private companies, the National Housing corporation, and other bodies. It is a forum for the different interests. But each prefecture and the Ministry of Construction can decide on its own. The institutes's role is coordination and harmonization, but it proves very difficult to get a plan: "A very Japanese style," we were told, since each prefecture tends to want to build its open empire.

This complexity reflects the fact that the basic initiative here – in contrast with Tsukuba – has come from the private sector. Big companies – Sumimoto, Matsushita, and Kawasaki Steel are among the biggest – have bought the land speculatively, thus overcoming the usual endemic problems of land assembly in Japan, and are investing their profits in it. They want to move into sunrise industries to replace the area's declining sunset industries, which have suffered as controlling influence has steadily passed to Tokyo since World War Two. They can claim special tax concessions for their investments.

The main reason for such an arrangement is the fiscal problem of government. In the early 1980s Japan experienced a fiscal reevaluation, the so-called *gyokaku*: now, it is difficult to provide 100 percent national funds for regional development. In fact both national and local government, in the cash-strapped 1980s, lacked resources and so see private investment as an opportunity. Incentives to the private sector – under a special law applying only to the Kansai Research City – include lower land costs, property taxes and accelerated depreciation. Tsukuba has the same kind of provision, and so do many other developments; each has a specific law, with subtle variations from the rest. The significance is that both Tsukuba and Kansai, which have the title of national projects, enjoy especially high levels of incentives specifically allowed by the Ministry of Finance.

In Kansai, the Association was founded earlier, in 1983, to conduct publicity and promote the area to outside industry. The Kansai Science City Promotion Act was passed in 1987 by the National Diet, establishing this as a national project, which gives special tax treatment; the construction plans of the prefectures were approved by the Prime Minister in 1988.

Already just completed, in summer 1990, is an extraordinary variety of different major projects, which made the whole region look like a vast construction site: the Advanced Telecommunications Research Institute International, Doshisha University, Doshisha Women's College, Doshisha

International High School, Osaka Electro-Communications University, Kansai Junior College of Foreign Languages, Kyoto Flower Center, and Kyoto Koseinenkin Resort Center. Under construction were the International Institute for Advanced Studies, Ion Engineering Research Institute, Hi-Touch Research Park, and Cultural Greenery Park. Still planned were the Advanced Institute of Science and Technology, Kansai National Diet Library, and the Kyoto University Faculty of Information Study. Among private companies who have already bought land are NTT, Sumimoto Metal Industries, Kyocera, CSK, Matsuhita Electrical Industrial Co., and Shimazu.

Almost needless to say, the plan depends upon completion of an ambitious infrastructure investment program, which was only just beginning to happen in 1990. There will be a direct connection to the new Kansai International Airport, the 40–mile (60–kilometer) journey taking one hour by a projected expressway which should be ready for the airport's opening in 1993. By 1991, there will be a direct expressway connection from Kyoto, plus a new Kyoto–Osaka expressway to be completed only by 2000; construction of the latter has been delayed because it goes through residential areas. These routes are nationally funded; a weakness of the plan is the delays in completion of infrastructure. Within 15 years, the Kintetsu private rail line will be extended from Osaka. A proposed Tokyo–Osaka maglev train route will also be extended from Kyoto to Nara with national support, perhaps with a station here; the Preparation Law of 1989 provides for a test line to check construction possibilities. The Ministry of Transportation loosely coordinates all these different initiatives.

We visited one recent arrival, the Ion Engineering Center Corporation. It carries out research on new ideas in ion engineering (ion-assisted technology). It is capable of controlling crystallization, components and various characteristics of materials, and will combine metals, semiconductors, ceramics and organics to create new materials. The basis of ion engineering is that a material is ionized either in a low vacuum or a high vacuum region. Kinetic energy given by an accelerating potential and/or an ionic change is used. The Center is introducing state-of-the-art equipment as an aid to R&D.

The Government persuaded the Institute to come here; the project was paid 33 percent by national funds, 17 percent privately; 50 percent was borrowed, to be paid back after profits come in. Thus the project is an entirely new national facility. It is managed by a management group with both government and private members, again, a new form of public–private operative management, which has arisen because of the financial problems recounted above. The origin lay in the private companies bringing in researchers. (Japanese professors may consult but do not work in developments like these.) The facility serves many different high-tech companies needing very large basic equipment – a scale economy. Very well-known companies like Toyota, Fujitsu, NEC, and Mitsubishi are represented.

Not far away, the Advanced Telecommunications Research Institute (ATRI) International exists to promote basic and creative telecommunications research and to encourage collaboration through joint research projects between industrial, academic, and governmental organizations, with a strong commitment to international exchange. Research includes such diverse aspects as automatic generation of communication software, combined speech and image understanding, manipulation in 3-D virtual space, automatic interpreting telephones, basic research in auditory and visual perception, and optical and radio communications. The complex is being built in the research core of the park.

Capital investment is subsidized to the extent of 70 percent by the national Government through the Japan Key Technology Center, backed by the prefectures. The main core research center was in Osaka for the first three years, but moved to Kansai in 1989. Researchers from many companies, 45 in all, are represented here, making 95 percent of the personnel. The Japan Key Technology Center comes from NTT stock sales five years ago, organized by the Ministry of Culture. The research is patented; users pay license fees. The research is, however, very basic and will not pay off in the short term.

Close by, the Communication Center of Culture, Science and Technology will be built in the Seika-Nishi-kizu district, the heart of what will effectively be a new cultural capital for Japan. Phase I construction started in 1990 and was due to be complete 1992; Phase II will follow from 1995 and the National Diet Library will open in 1996. The Center will provide integrated facilities for advancing knowledge exchanges through a conference center where overseas visitors will meet their own staff; it will serve as a liaison for cooperative research projects in unusual fields or enterprises, with low-cost access to information; it will develop seminars for business leaders and citizens, addressed by visitors to the ATR, the Ion Engineering Research Institute, the Advanced Institute of Science and Technology and other institutions; and it aims to further cultural and scientific exchange through a sophisticated information system.

Not far away, the Touch-1 Hi-Touch Research Park is a privately-funded activity with a number of leading companies in the field of consumer electronics and the media, each with its own R&D facility, dealing with such diverse aspects as electronic entertainment, "smart" homes, the tea industry, and even wedding ceremonies.

The International Institute for Advanced Studies, a nonprofit research foundation, established in 1984 with strong industry–government–academic support, will occupy its purpose-built facilities in 1992. It will limit itself to basic studies on human subjects, will be open to scholars from all over the world, and is based on flexibility of research topics. In the "Scholars'

Village," scholars will be invited for between a few weeks and one year, and will research on their own topics, attending workshops to exchange work. The proposed research areas are very general: "Creation of New Culture," "Social System for the 21st Century," "Principles of Natural Evolution," "Theoretical Life Science," "Thoughts and Actions," and "Safety Science." Much of the activity so far seems to have been in only two fields, biotechnology and psychological perception, though this may be changing.

Overall, the Kansai Research Park bears all the signs of an outstandingly successful program, backed by massive national support. The irony, as with Kanagawa/Kawasaki (Chapter 7), is that it flies in the face of MITI policies to divert such developments out of the Tokyo and Osaka metropolitan areas. Here, as elsewhere, it seems that Japan's national technology development objectives and regional development objectives come into head-on opposition. But Japanese government policies have proved to be amazingly capable of accommodating such contradictions, and will doubtless do so here.

CONCLUSION

The four important experiences presented here provide some clues that help shape answers to the fundamental questions underlying our inquiry, as formulated at the start of this chapter. First, it appears that spatial concentration of research activities has little effect on scientific innovation in the absence of a deliberate program to favor synergy, and of specific mechanisms to implement such a program. If there is no system to stimulate networking and cross-fertilization, the science units located in the new area are only as valuable as the value of their individual members. If there is scientific excellence in the institutes established in the science city, research findings will undoubtedly emerge, but these will not necessarily be more far-reaching or more industrially applicable than if they had originated in a more traditional academic setting.

Indeed, in some instances, the fact of moving the research units from their original networks may hamper the quality of the research, by increasing their isolation *vis-à-vis* national and international scientific milieux. True, the improved working conditions and scientific equipment, which generally go hand in hand with the investments in establishing the science city, may help research activity. But such improvements might yield even more productive results if provided to existing units of academic excellence. The organization of new research institutions may have a positive effect in breaking bureaucratic rules and in undermining the conservative ideology of established academic centers. Yet, unless new systems of research management and organization are established, our observation shows that, after a short period of time, old scientific vices are simply reproduced in the new science cities. In such conditions, research institutions, researchers, and research directors all carry with them the viruses of their own demise. By relocating without reorganizing, the old problems will not fade away.

The experience of Akademgorodok is telling here: it shows the total lack of scientific value added that comes from the spatial concentration of research institutions in a new, isolated setting. Of all the cases we studied, Akademgorodok probably had the highest scientific excellence at its onset. And yet, its isolation from both the Soviet Union and from the West, the lack of cooperative mechanisms among the researchers, the vertical organization of the Academy of Sciences under the conditions of a Central Plan, and the complete isolation of the scientific work from the economy and from the society, led to the decline of scientific productivity and of scientific excellence in the institutes located in this Siberian Utopia: the once-young, innovative researchers reproduced in their older years the same bureaucratic behavior that had led them to escape from the heavy hand of Moscow.

Second, it would seem that the development of a science city requires a fundamental impetus from the public sector: it must be anchored from the outset on the location of government and/or university research institutes. It is extremely difficult for private firms to take the initiative in such a risky venture without a strong commitment from government agencies, including the presence of government research institutes that provide a basis for reaching the critical mass of researchers and research activities in a given location. But the more a technopole is based on public research institutes, the more it is difficult to link up with industrial applications that make the research economically useful. While such linkages are still possible, they require the expansion of the original nucleus of public research institutes into the domain of private firms' research institutes, which alone will provide the channel for the commercialization of research findings. The two-stage process in Tsukuba illustrates this analysis, with the private firms being called in by the Government itself in the 1980s, after the initial failure of the public research institutes to generate sufficient research with direct industrial and agricultural applications. Thus, if the goal of a science city is to provide the research basis for economic productivity and competitiveness, it seems that the integration of public agencies and private firms at the very onset of the design is a precondition for future linkages between basic research, R&D, and industrial applications – as appears to be the case in Kansai.

Third, it does not seem that in themselves science cities are powerful tools of regional development, at least in the absence of specific policies linking the information they generate to the local and regional economy. Both Akademgorodok and Taedok have remained exotic enclaves in their regional environment, actually condemned by local authorities as a waste of scarce resources, unconnected in any way to the regional economy, generating few jobs, the most skilled of which go to scientists and technicians from other areas. Yet, if the conditions of regional development change, if the areas where science cities are located become dynamic in their own right, they can then make use of the scientific potential contained in their science cities. Thus, when the expansion of metropolitan Tokyo finally reached Tsukuba, this science city suddenly became an additional element of the multinuclear technopole that is emerging in Greater Tokyo. If the current tendencies of articulated regional development in South Korea end by integrating Taejon and the central area of the country, Taedok will certainly be an asset for the region. Kansai is already playing a role in the reindustrialization of the Kyoto–Osaka area, actually increasing metropolitan concentration rather than fostering regional decentralization. And Akademgorodok is showing new signs of vitality in the 1990s, as its scientists start working as consultants and experts for Siberian enterprises engaged in a process of modernization, to open up to the international economy.

Thus, undoubtedly, the presence of thousands of scientists and engineers in an area is a fundamental asset for the economic dynamism of that area in an information-based economy. But the science city will be directly productive for that regional economy only as long as it is materially related to its productive activities, through the integration of its knowledge and expertise into a network of regionally-based enterprises.

It would seem, though, as if the efforts involved in designing and constructing science cities are too great in comparison with the value added they generate, be it in scientific excellence, in industrial applications, or in regional growth. Under such circumstances, why build them? Clearly, countries that have low scientific potential, or whose scientific institutions seem to be paralyzed by bureaucracy, can profit from a clean start in the form of new institutions, which concentrate human and financial resources in a new setting, and which cut themselves off from the past. Yet, most of the cases of science cities in the world, and three of the four cases studied in this chapter, seem to stem mainly from the internal logic of the State. It is the creation of a new space, through a scientific project designed and controlled by the State, that seems to be the fundamental goal of science city projects.

The relevance of such an enterprise can be better understood by considering the symbolic and material importance of science, and of the appropriation of science, in the new technological paradigm we inhabit today. By establishing a new center of science under its direct control, the State declares its capacity to master modernity and power simultaneously, breaking away from the old molds to shape the future through the instruments of science. Science cities are thus generally born out of the messianic dreams of developmental states. Once they exist, their scientists, institutions, organizations, and dwellers muddle through the reality of their regional environment; they try to reach more modest goals of generating knowledge and selling it in the local markets. While never quite attaining the scientific utopia of their authoritarian creators, science cities do become the quarries for the raw material on which regional futures are increasingly based: newly-minted information.

5

TECHNOLOGY PARKS: INDUCING THE NEW INDUSTRIAL SPACE

Technology parks have become a fashionable policy in local and regional economic development. They aim to concentrate in a designated area a number of high-technology industrial firms that will provide jobs and skills, and eventually will generate enough income and demand to sustain economic growth, in a region that is seeking to survive in the new conditions of international competitiveness and information-based production. Their emphasis is on manufacturing, although some specialize in the R&D component of manufacturing.

Governments, be they national, regional or local, tend to play a decisive role in the design and development of technology parks, but the universal goal of the projects is to attract investment by private firms. To succeed in attracting the best firms in a context of competition between localities around the world, governments use fiscal incentives, offer facilities and productive infrastructure, accommodate specific demands from incoming firms, seek to improve telecommunications and transportation, help the creation of educational and training institutions, and work hard to create a favorable image for the park, usually by improving the environment and staging public relations campaigns. While the technological level of a firm is often an important criterion for admission into the park, the standards by which the park's success are judged may be rather different in practice: job creation, both in quality and quantity, and the importance of investment loom large.

Ultimately, technology parks more closely resemble new-style industrial districts than innovative milieux. The main concern of the developers of a technology park, whether public or private, is to generate a growth pole that will attract further investment by firms in high-growth industrial sectors. The assumption is that the higher the technological level of the firm, the more it will grow and compete in the new economy. But it is industrial competitiveness, rather than scientific quality, that remains the fundamental goal of any technology park project.

Technology parks are a very widespread phenomenon worldwide; it seems that hardly a region, scarcely any self-respecting city, lacks one. We will analyze the issues that they raise by concentrating on three different cases that represent different contexts and different industrial structures. Sophia-Antipolis, on the French Riviera, is one of the most highly-publicized technology parks, which combines the European branch establishments of multinational corporations, small and medium firms, and large public research centers and universities, under the auspices of public regional authorities. The "Cambridge phenomenon," built around the Cambridge science park in England, is a semi-spontaneous spin-off from a major research university, which gave birth to a technology-oriented complex consisting essentially of small and medium firms. The Hsinchu Science-Based Industrial Park in Taiwan illustrates a more deliberate national Government project to attract advanced foreign firms into a new area, built around national research centers and universities, in order to diffuse technology and industrial know-how into networks of local firms, as a way of upgrading the Taiwanese industrial structure.

SOPHIA–ANTIPOLIS

The French media, and the local public relations industry, are in no doubt: Sophia-Antipolis, the international business park near Nice on the Côte d'Azur, is a classic example of a "technopolis." Glossy brochures, which greet the visitor at the park's reception desk, advertise "The European Smart Site" ("*Le Site Intelligent d'Europe*") and "A Technopolis for the 21st century . . . Under the sunny skies of the French Riviera, between Nice and Cannes, a Technopolis for the whole of Europe."[1] And certainly, at first sight the achievement is impressive enough: 400 companies and 9,000 workers on a 5,250-acre (2,300-hectare) site, in a unique landscaped park, with a promise of 4,000 additional acres (1,600 hectares) and 25,000 more jobs by the year 2000; all served by an advanced telecommunications network, and located a mere 11 miles (18 kilometers), from France's second international airport (Figure 5.1).[2]

Most remarkably, and unusually among the schemes considered in this book, it was in origin the notion of an individual, which was fully accepted as a public initiative only after a decade of indecision. It was the brainchild, in 1968, of Pierre Lafitte: then deputy director of the Ecole Nationale Supérieure des Mines de Paris, now Senator for the Alpes-Maritimes. He had a remarkable vision: a city of science, culture and wisdom, to be created on the plateau of Valbonne – the only significant developable space on the entire Côte d'Azur, which up to then had lacked access and basic services. The decision to develop was taken a year later; by 1972, consultants had developed a land-use plan that reserved one-third of the site for a mixture of innovative technology, housing and "living activities," leaving two-thirds as green belt, with the hilltops remaining open. This strict respect for the environment gave the entire development a very high quality. That same

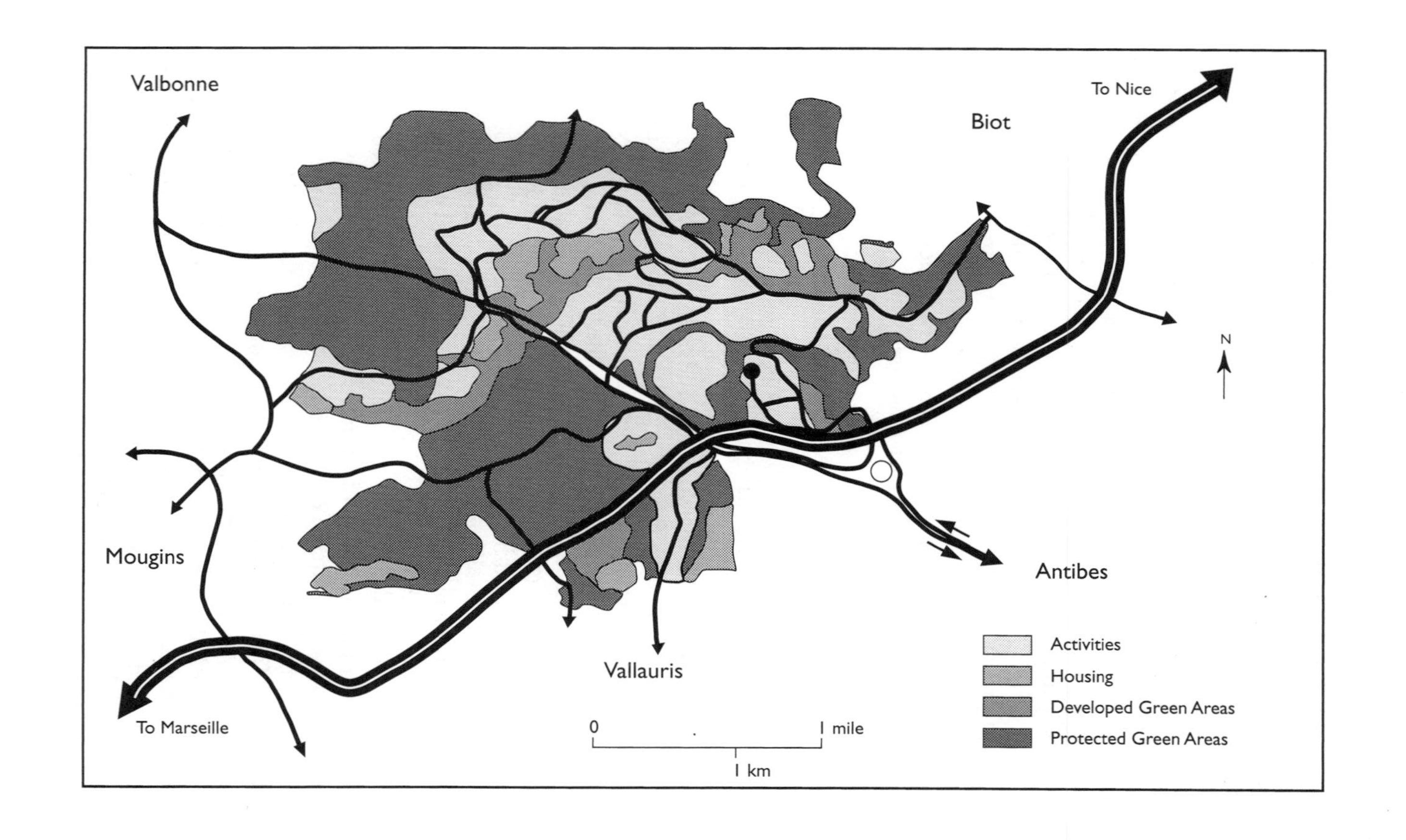
Valbonne
To Nice
Biot
N
Mougins
Antibes
Vallauris
To Marseille
0
1 mile
1 km
Activities
Housing
Developed Green Areas
Protected Green Areas

year, 1972, the first institutions came on to the site – among them, the decentralization of Lafitte's Ecole Nationale Supérieure des Mines de Paris (ENSMP), which was to play a critical role in the success of the scheme.[3]

But, during much of the 1970s, progress was slow; Sophia-Antipolis was seen as a kind of high-technology island, divorced from the rest of the region. In 1972, a National Interministerial Committee approved the plan, which involved diversification of the economy, creation of employment, and concern for the environment; it declared Sophia-Antipolis as of "National Interest," a designation that meant easier funding; and it created an interministerial group to coordinate activities. But, without specific government backing, the first stage involved great risk; DATAR, the French national agency for regional planning and development, was in favor but other ministries were reserved if not hostile, and funding actually dried up at one stage.[4]

Locally, a syndicate, Symival, was set up in 1975 to represent all five local communes, the Department and the local Chamber of Commerce and Industry for Nice–Côte d'Azur; it began to develop the infrastructure and to delegate powers of commercial development. Finally, in 1977, the public sector entered the stage and the park took off; the enterprise was powerfully assisted, at this time, by the creation of a semi-public company to try to draw foreign investment outside the Paris area. Almost simultaneously, Air France brought its worldwide passenger reservations service here, reassured by the presence of IBM and Texas Instruments; this in turn persuaded Digital. The years 1977 to 1981 were a time of rapid employment growth. In 1982, François Mitterrand's administrative reforms gave more power to the local *département* of Alpes-Maritimes, which has general administrative responsibility for the site and is partly responsible for management through a semi-public management company, SIAM. The *Contrat de Plan* for the 1984–8 four-year plan at last recognized Sophia-Antipolis as an element of regional development; but still rather as part of a policy of deconcentration of public enterprises and services to a "French Sunbelt," than as part of a truly synergistic development process.[5] Investment rose sevenfold between 1982 and 1989; at last, Sophia-Antipolis acquired its national and international reputation as a technopolis.[6]

Already, by the mid-1980s, total investments were some FF6,000 million in 1985 prices, with a direct annual turnover of FF1,200 million.[7] The scheme had been (and is) financed up to 70 percent by land sales, so there was not much burden on the local communities who enjoy the business tax. Originally the companies were in charge of their own infrastructure but did not pay taxes, hence the finance through land sales; management of roads, woodlands and equipment is now controlled by the communes, but in reality Sophia-Antipolis has been the partner of the companies, who have formed a Managers' Club and pressure Sophia-Antipolis to maintain the quality. Incentives take the form of equipment – fiber optic cable for telecommunications, and an advanced telecommunications research center – rather than financial aid; but firms enjoy reductions in business taxes, which tend to be high on the Côte d'Azur, in comparison to the national average.

Sophia-Antipolis at the end of the 1980s

By 1989, Sophia-Antipolis housed some 400 companies employing some 9,500 workers; about 70 percent could be regarded as high-tech.[8] The park has a strongly international character: most companies bring in engineers from all parts of the world. Thus at Digital, with 900 employees, 40 percent are foreign; 25 nationalities are represented. In addition, the park housed some 6,000 residents, the great majority of whom worked within the park. But only 10 percent of the workforce lived here; most people commuted in. There was a small town with 1,300 homes built between 1981 and 1988, plus 200 shops, service agencies, professional consultants, associations and clubs, as well as two school centers for infant–primary education, and a high school.

The economic structure is dominated by a few major companies such as Digital (870 employees), Thomson Sintra (473), Air France (330) and Télémechanique Electrique (306), which however live side-by-side with a great number of much smaller firms (Tables 5.1 and 5.2). The group comprising computer sciences, electronics, robotics and telecommunications, which employs half the workforce, includes such major presences as Digital, Aisin Seiki, INRIA, Amadeus Development, ETSI (the European Telecommunications Standards Institute), Air France reservations, Télésystèmes, Rockwell International, Télémechanique, VLSI, Phoenix Technology, Thomson-Sintra, France Télécom and SITA. The second biggest cluster of employment – embracing medical sciences, chemistry, biology, and accounting for more than one in eight of all workers – includes Dow France, Dow Corning, Cordius, Rohm & Haas, Searle, Wellcome, CIRS (International Center for Dermatological Research), and the new Laboratory

Table 5.1 Sophia-Antipolis: structure of employment (January 1989)

Total	9,500
Information, electronics, telecommunications	4,058
Health sciences	1,107
Energy	299
Higher education and research	642
Commerce	184
"Liberal professions"	109
Associations, committees, foundations, clubs	41

Source: Sophia-Antipolis, 1989: 1
Note: "Higher education and research" includes only those in research institutions, and excludes those in industrial laboratories

Table 5.2 Sophia-Antipolis: analysis of firms

Type of industry	*Number of employees*			
	<10	*10–99*	*100+*	*Not stated*
Information, telecommunications, electronics	45	39	10	27
Health sciences	6	4	6	5
Energy	8	7	0	3
Higher education, research	7	8	2	9

Source: Société Anonyme Sophia-Antipolis, data on industrial installations

for Molecular Biology, a joint enterprise of CNRS and the University of Nice.[9]

Research, education and training activities employ about 2,500 out of the total of 9,500: about one-quarter of both employment and annual revenue.[10] They include the Ecole Nationale Supérieure des Mines, one of the first arrivals; the University of Nice; and Advanced Training Colleges such as ESSTIN, CNAM or CERICS. In addition the CERAM group, created by the Chamber of Commerce and Industry for Nice–Côte d'Azur, offers a range of Master's degrees in association with Nice's Schools of Advanced Commercial Studies. The list of higher education institutions is an impressive one: the Université de Nice–Sophia-Antipolis, Institut Universitaire de Technologie, Ecole Supérieure en Sciences Informatiques, Ecole Supérieure des Sciences et Techniques de l'Ingénieur de Nancy, Institut Supérieur d'Informatique et d'Automatique, Centre d'Enseignement et de Recherche Industries et Sciences de l'Informatique, Ecole Nationale Supérieure des Mines de Paris, Centre d'Enseignement et de Recherche Appliqués au Management, Institut Français d'Ingéniérie du CNAM, and Centre de Formation Internationale à la Gestion des Ressources en Eau. Natural sciences, particularly concerned with energy and new materials, form a fourth area including such employers as AFME, KIER and CNRS.[11]

By the beginning of the 1990s, a 5,000-acre (2,000-hectare) expansion was planned to the north; the notion was that progressively, the park would become a self-generating system, spreading into the surrounding area and eventually developing as a linear "Sun Belt" from Barcelona to Turin.[12] Already, within the local *département* of Alpes-Maritimes, there are many other high value-added industries like IBM's Study and Research Center at La Gaude, Aérospatiale, and TI's semiconductor plant and European HQ at Villeneuve-Loubet. It is claimed that the existence of Sophia-Antipolis has been directly responsible for the creation of 30,000 jobs in the *département*, and that 100,000 jobs – 10 percent of the entire labor force of Alpes-Maritimes – are connected in one way or another with the park. Total turnover from industrial and commercial activities now accounts for 24 percent of all turnover, with the largest share coming from Sophia-Antipolis.

Evaluating the Sophia-Antipolis achievement

There have been two extended academic evaluations of Sophia-Antipolis, by Perrin (1986a, 1986b) and Quéré (1990). Both reach somewhat pessimistic conclusions, somewhat at variance with the media or public relations pictures. It is interesting to ask why.

Perrin shows that between 1974 and 1986, the park attracted 5,700 employees, of whom 2,860 (47 percent) were in high-technology industries on the most rigorous definition; adding routine management activities and services, the total was 3,500, or 60 percent. Sophia-Antipolis, he found, has a dual structure: it is both a *parc de prestige* for established multinationals, and also an incubation center, both in basic research and in management, for small and medium enterprises (SMEs). The two structures are very different. The first attracts established firms by imitation, reinforcement, and image; it has provided most of the jobs. The second, in contrast, depends on "collective apprenticeship" through communication and through launching new projects; it has not been very successful in generating a critical mass of new small and medium enterprises.[13]

Perrin suggests that the dual structure may have actually inhibited the necessary cross-fertilization, both vertically between R&D and production, and horizontally between sectors. Experience in both Silicon Valley and Cambridge, he argues, shows that the process of incubation takes a long time. Regionally within the Côte d'Azur, the high-tech industrial tissue is weakly developed; there is a weak system of subcontracting, so that small and medium enterprises are underdeveloped; so they then find that their costs – for land, maintenance, qualified labor – are too high; and they lack basic services, both technological (testing laboratories) and organizational (risk capital, marketing). And, since the large firms provide generous pay and facilities for their researchers, the latter never develop the inclination to break away and start their own firms; their employers continue to internalize their R&D. Thus, though the structures of incubation are in place through the organs of research and education, their operation is hobbled by the structures that stem from the global–national environment.[14]

Perrin thinks that there may be a way of escaping from this vicious circle. It is a promising sign that Sophia-Antipolis has developed a common scientific culture, a common concern with research, a common focus on innovation. Established firms, he argues, now need to manage innovation through agreements with universities, laboratories and other sellers of technology. This could be encouraged through joint ventures between existing laboratories and private firms, and the creation of some central service of advice to SMEs through the existing Centre d'Etudes et de Recherches en Management.[15] This last notion seems to have been fulfilled by the establishment – at the beginning of 1985, with the aid of the *département* of Alpes-Maritimes and the Sophia-Antipolis Foundation – of a *pépenière*, or business incubator: the Centre d'Acceuil des Technologies de Sophia-Antipolis (CAT) offers young entrepreneurs administrative services, relations with a network of scientific, technical, commercial and administrative specialists, and availability of small lots.[16]

A more detailed analysis comes from Michel Quéré.[17] He argues that Sophia-Antipolis, together with the development around the city of Grenoble, is the only true example of successful technological development in France. But Sophia-Antipolis actually consists of three "poles" which remain quite distinct: Information-Telecommunications-Electronics (ITE), Pharmacology-Biology-Chemistry (PCB), and Energy, which has stagnated and remains far smaller than the other two. The two bigger poles are quite different: ITE has a multitude of small firms clustering around the bigger one, while PCB exhibits no such effect.

Quéré is concerned to ask where, in an area like Sophia-Antipolis, we find an "innovative milieu": an organization that develops new ideas itself, in advance of the market, perhaps through cooperation with other local firms. He finds an important difference between outside and indigenous firms (Table 5.3). Very few of the outside firms represent an innovative milieu, either in whole or in part; most of the indigenous firms on the other hand can be described as innovative, but not all.

Table 5.3 The "innovative milieu" at Sophia-Antipolis

Part of the "innovative milieu"
Outside firms that are developing technologies through local external relations
Outside firms that have located many small units
Local spin-off firms
Local firms developing through subcontracting to outsiders
Local firms working through associations with research centers
Not part of the "innovative milieu"
Local outside firms, which existed prior to the park's establishment
Outside firms that have decentralized part of their activities
Outside firms that have decentralized their R&D
Local firms serving as interfaces between producers and consumers
Local firms providing technological services

Source: Quéré, 1990

Quéré uses his categorization to show that in the case of both the ITE and the PCB poles, the creation of a true technopolis came relatively late: in the case of ITE from 1979, in the case of PCB from 1980. They appear, as he puts it, only in a stage of "maturity of the productive space."[18] That stage saw the arrival of certain major outside firms which, though not locating in the park in order to draw on the resources of the area, nevertheless encouraged the development of complementary services. For instance, Air France reservations, which was here earlier, began to draw on local resources to deal with its problems in real time.

The "innovative milieu" of Sophia-Antipolis thus consists of two relatively distinct classes of firms. First, there are outside firms which either organize themselves in several units or entities operating as quasi-firms and developing partnerships, and those that seek related expertise and develop a "common apprenticeship." Second, there are local firms, which tend to be narrowly related to research centers and to develop more specific technologies. But the entire technopolis has failed to develop to its true potential, for three reasons.

The first is that the firms constituting the milieu possess expertise in quite distinctly different segments, a fact that does not favor partnerships or the development of cross-cutting activities. The second, which logically follows, is an isolation of the human resources that must be the basis of any milieu of innovation; analysis of the labor market shows that there is virtually none of the mobility between firms that is so characteristic of Silicon Valley. The third is that firms and research centers see an actual danger in cooperation. That especially applies to the outside firms, who tend to stand at arm's length from the national research centers; perversely, proximity comes to play a negative role. An analysis of relationships shows clearly that these outside firms remain isolated; only the indigenous firms tend to form partnerships with the research centers.

Quéré finally suggests some policies that might help: encourage local agreements between firms which could develop vertical integration; reinforce the local higher educational infrastructure; develop local risk capital; and broaden the management structure to include universities, research centers and firms.

Summing up

Though Sophia-Antipolis was originally a private venture, it was effectively taken over by the State in the late 1970s. And herein lies an important general conclusion. For, as Quéré emphasizes, France has a long tradition of centralized administration. When the Decentralization Law of 1982 gave new powers to the *départements*, the notion of developing "technopoles" on the model of Sophia-Antipolis became suddenly very attractive, resulting in more than 40 developments of this character during the 1980s. Yet the tradition remained centralized; it was merely centralized at the level of the *département* rather than at that of the nation. So the State has continued to take the initiative, while the other actors, private and public – the universities, the enterprises, and the research centers, which must be the driving forces behind any such development – have remained subsidiary, deprived of any real power over directions or decisions. The dominant public actors have seen the issue more as one of land-use planning than of economic development: success has been measured in sales of so many square meters of land, an obsession with the packaging rather than with the contents of the package.

It follows, Quéré argues, that the public authorities are incapable of the kind of technological management that is needed. Sophia-Antipolis, like Grenoble, became such a media success in France because it seemed to be different. But in reality it was not: there remained two distinct entities, the big state research centers and the multinationals, that did not effectively relate to one another because their worlds were too closed.

One is irresistibly reminded of Perrin's judgment on the *Cité Scientifique Ile de France Sud*:[19] that, in Europe's largest agglomeration of high-technology research and development, the synergies and cross-fertilization remain minimal. "Each unit," he concludes, "behaves like a citadel jealous of its power and of its supposed superiority."[20] And he quotes a conclusion by two other researchers, Chambon and Dyon: "Research and industry regard each other like china dogs on the mantelpiece and reproach each other for their limitations." The same phenomenon, Perrin notes, has been observed in Tsukuba. And, evidently, it equally applies to Sophia-Antipolis.

Admittedly, Sophia-Antipolis is a success at one level because it has worked: the park has been developed, it has brought in firms and jobs. But it has not so far worked at the deeper and more critical level, which is the creation of a true milieu of innovation: the necessary synergies are not yet richly developed. It may be a matter of time, more time perhaps than the two decades, so far, of life at Sophia-Antipolis: time, indeed, is the problem with such *ex nihilo* creations as Sophia-Antipolis. The verdict, as with other such developments, needs to be suspended.

CAMBRIDGE

In the mythology of high-technology creation, the "Cambridge phenomenon" has come to parallel Silicon Valley; it has become a worldwide image or symbol of the innovative milieu. The term itself was a happy invention of local consultants, in a research study made in the mid-1980s,[21] which immediately fired public and media imaginations. And the "phenomenon" to which it referred is a much more recent one than Britain's M4 Corridor (or Western Crescent). While the Corridor-Crescent was already well-established by the 1960s, Cambridge as a high-tech center is essentially a creation of the 1970s and 1980s. Further (and this, without doubt, was the fact that made the report so exciting), Cambridge appeared to represent genuinely entrepreneurial, new-firm-based growth – based on computing hardware and software, scientific instruments and electronics, and, increasingly, biotechnology – which had spun off from university research on the Silicon Valley model.

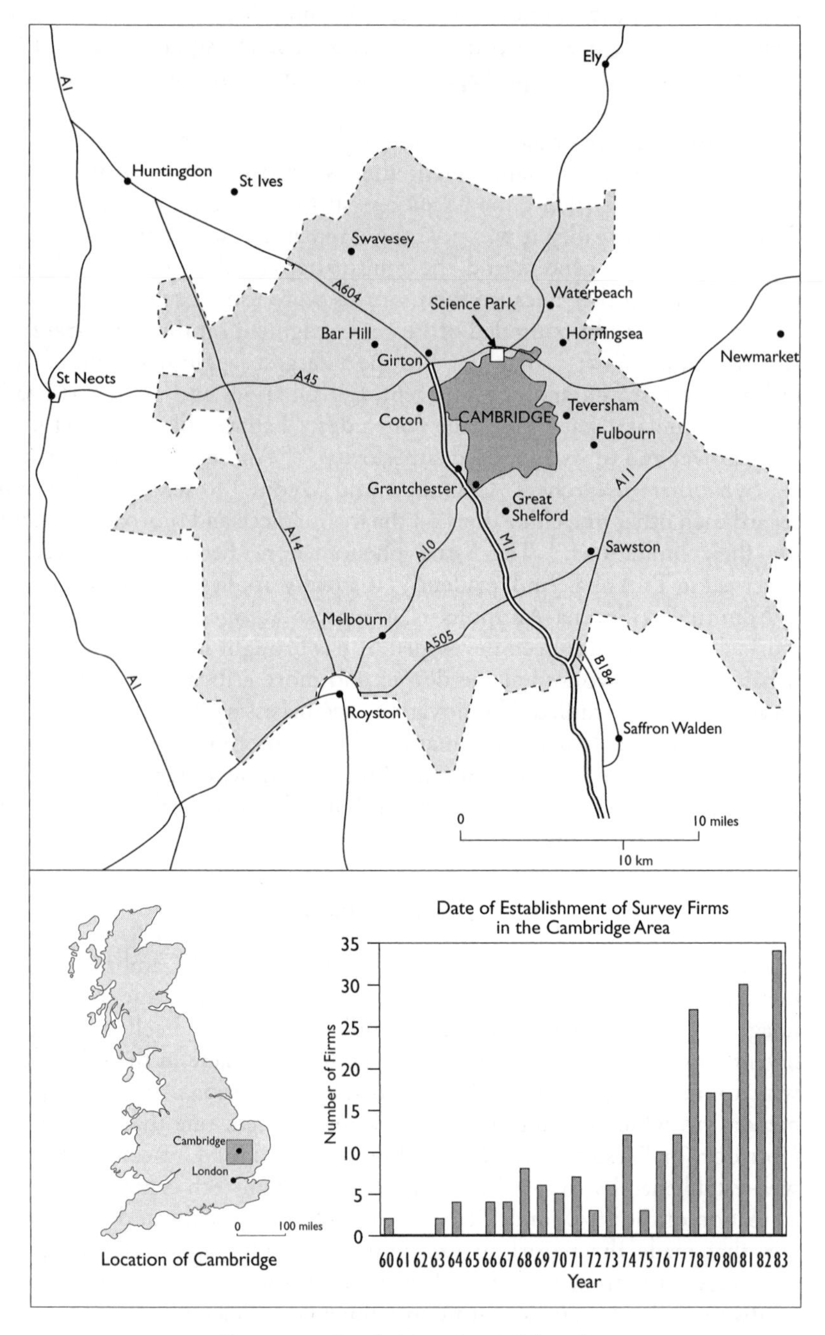

Figure 5.2 Cambridge: general location
Source: Segal Quince Wicksteed, 1985

What made it additionally interesting was that it had occurred in a rather unlikely place. Cambridge lies 50 miles (80 kilometers) north-north-east of London, in East Anglia: a region that, as late as the 1960s, was generally regarded as a rural backwater, but that suddenly took off to become the fastest-growing region of Britain (Figure 5.2). During the early 1980s, the county of Cambridgeshire recorded the largest volume and the third fastest rate of high-technology employment growth in Britain. By January 1986 some 16,500 workers were found in some 350 high-tech firms in the southern part of the county, around the city of Cambridge, providing more than 11 percent of total employment. Research by Cambridge University's geography department showed that it constituted a genuinely "technology-oriented complex" characterized by processes of synergistic interaction between new firms, banks and finance agencies, and business service organizations, and underpinned by research institutions and an attractive environment.[22]

These firms are in R&D rather than production: no less than 42 percent described R&D as their main activity. Their effect on employment generation has been fairly modest – a net growth of 6,000 jobs, or 63 percent, in the Cambridge region between 1979 and 1987 – but appears more impressive against a total employment of only 120,000, and a national manufacturing job decline of some 27 percent in the same period. Further, these are high-income jobs, and the total indirect job creation may be as large again.[23]

The Segal Quince Wicksteed analysis shows however that the Cambridge phenomenon did not suddenly appear out of the blue; there were perfectly good reasons why it should have happened here. Cambridge had an early tradition of high-tech manufacturing through spin-off from university laboratories: Cambridge Instruments had been founded as early as 1881 to manufacture scientific equipment for the University, while W.G. Pye left the Cavendish Laboratories to set up his own firm in 1896. Some other small scientific companies were set up in the 1930s, including the precursors of Fisons.[24]

These might have been expected to generate expansion after World War Two, because the University took an early lead in computer development. But planning policies intervened. An official report by an eminent planner, Sir William Holford, published in 1950, recommended that, to preserve the university city's character, it should be physically constrained from growing and that new industry should be discouraged from locating here. Accepted by government, it guided planning policies for the next two decades, causing a very negative attitude to development. Locally-based companies were compelled to relocate outside the city when they were refused permission to expand inside it; IBM was refused permission to set up its European R&D HQ there.[25]

A critical turning-point came in 1967, when the university set up a Senate subcommittee to "consider in greater detail and advise on the planning aspects of the relationship between the University and science-based industry"; its report, named for its chairman Professor Sir Nevill Mott (head of the Cavendish Laboratory) was published in 1969. It concluded that the Holford principles had created problems both for the University and for industry; it recommended limited growth of existing and new science-based industry and other applied research units in and near Cambridge, and the establishment of a "science park." In 1971 Cambridgeshire revised its development plan accordingly – a step that was helped by a change in the geography of the county, which incorporated the high-unemployment Peterborough industrial area at this time. Thus the Mott Report was a turning-point that led directly to the establishment of the science park, and established the fact that science-based – as against smokestack – industry was acceptable in Cambridge.[26]

The effect was explosive. In 1959 there had been 30 high-technology firms in the Cambridge area; by 1974 there were 100, by 1984 no fewer than 322, accounting for 17 percent of employment in the Cambridge travel-to-work area (Figure 5.2). New-firm formation averaged one per month between 1964 and 1984, and 1.5 per month during 1974–84, a rate that almost compares with the far bigger San Francisco peninsula. Total employment by 1984 was estimated at 13,700 and output at £890 million, of which 3,800 (£350 million) came from firms established in the previous decade. Between 1979 and 1984 aggregate employment increased by some 4,100 or 43 percent, of which 90 percent came from new firms established after 1974. Twenty-two percent of firms, with 21 percent of employment, were in electronics capital goods, 23 percent (8 percent of employment) in software, and 17 percent (22 percent of employment) in instrument engineering.

The great bulk of these companies were small, with a median size of only 11 employees. They were concentrated in the Cambridge sub-area, especially in Cambridge itself. A "family tree" shows that though only a small proportion, 17 percent, of companies was formed by people coming straight from the University, in indirect terms the University has been the origin of virtually all the companies, either because of indirect spin-offs or because the existence of the University was the reason for start-up.[27]

No less than 73 percent of firms interviewed by Segal Quince Wicksteed said they had started here because the founder lived here. Some 30 percent of their workers were scientists and engineers; and 27 percent of their graduates came from Cambridge. Slightly over half the firms had links with local research, of which 90 percent were with the University. They did not have important links otherwise with the local economy, but they had enjoyed access to start-up capital and management advice from the local branch of Barclay's, one of the British retail banks.

Segal Quince Wicksteed's analysis examines three possible factors which might explain the Cambridge phenomenon:

1 *Demand* played a role: in some fields (CAD, scanning electron microscopy), Cambridge was in a favored position because it was developing products for which there was a rapidly rising demand.
2 *General preconditioning*. The general growth of East Anglia and the favorable business climate for small firms came concurrently with major technical advances in electronics and computer design. The university encouraged research excellence; it enjoyed generous research funding, including major Research Council units. It has combined critical mass with quality: Trinity College alone had won more Nobel Prizes for science than France. It has exploited research in areas where it has done particularly well, and where start-up costs are low. The college-based structure of the University weakened departmental hierarchies, encouraging individual flair; and the University had an extremely permissive attitude to intellectual property rights, which belong to the researcher. And after 1969 it positively encouraged industrial spin-off. Culturally, the city was small, dominated by the University, and free of old industrial structures; new firms did not feel "lost" here. The Trinity College science park was very visible and was perhaps the most successful university-based science park in Europe; though it played only a minor role in the entire phenomenon, it had become increasingly important because of its policy of offering good premises on short terms at reasonable rents.[28]
3 *Special factors*. There were a number of key triggering events: the establishment of Cambridge Consultants in 1960, and of Applied Research of Cambridge in 1969; the Mott Report of 1969; the formation of the Cambridge computer group in 1979; and the lending policies of Barclay's Bank from the late 1970s. There were many young people who could not find university posts, but wanted to stay here.[29]

Many of these features help explain why the "phenomenon" should have happened in Cambridge and not somewhere else. The timing results from the interplay of five developments that happily came almost simultaneously: the Mott Report, the ripeness of technologies, new roads (the M11 from London, completed in 1979), the responsiveness of the financial and business services sector, and a cumulative demonstration effect. The large number of spin-offs was a product of several factors: the nature of the technologies, which encouraged small firms; the individualism and high quality of the people; the presence of Barclay's Bank; the lack of alternative employment; the relative lack of existing companies capable of using the knowledge; the habit of people leaving to start up their own companies; and, again, the demonstration effect.

The local environment is very attractive: a recent study by David Keeble (1989) found that 70 percent of high-tech entrepreneurs were immigrants, and that 79 percent of these had been greatly influenced by their environmental perception. And the city is close to London for capital markets, customers, suppliers, and international communications. The negative side is that, partly because the city has tried to restrict growth, house prices have risen faster than anywhere in the country except the South-East; yet housing growth may reduce the very attractiveness on which the region depends. And the phenomenon might just be on a course of self-destruction through the takeover of small companies by multinationals, as of Acorn by Olivetti or Applied Research of Cambridge by McDonnell Douglas.[30]

That last conclusion echoed an analysis by AnnaLee Saxenian, which suggested that the entire phenomenon may have been illusory, like the grin on the face of the Cheshire Cat in Carroll's *Alice in Wonderland*. Re-examining the record, she found a "lackluster performance": most of the employment had been created before the "high-tech boom" of the 1980s, which could come to an abrupt end (a prediction that proved all too accurate); most new firms remained very small, they got no help from larger companies or from government procurement, they did not network much with each other, and takeover activity was on the rise.[31]

Lessons from the Cambridge phenomenon

What lessons, then, does Cambridge offer? David Keeble believes that because the Cambridge phenomenon was not deliberately planned, its experience may not offer any obvious lessons to other places. Though the University's role was important, that might not be true elsewhere. The local environmental quality, especially the small-town atmosphere, seems to have been significant but may not easily prove replicable. Government policies seem to have been irrelevant, since no help was available. Maybe, Keeble concludes, it will be important to aid threshold firms that have grown to a significant size, and that then become threatened by takeover bids. Finally, the Cambridge story indicates that it is possible to start from virtually nothing and to achieve a critical mass, but that this takes time.[32]

This is underlined by the experience of science parks created around other British universities. In a follow-up study, Segal Quince Wicksteed compared Cambridge with three others – Warwick, Salford and Newcastle. They found that while Cambridge benefited from its small-town environment, Newcastle and to a lesser extent Salford were handicapped by the legacy of early industrialization. Warwick, a greenfield site on the edge of a major industrial conurbation, had done better. Warwick like Cambridge had also gained from networking among key individuals. Here, because Warwick saw the need earlier and more clearly than any other British university, the science park soon became a leading element in the relationship between the university and local industry. Warwick and Salford both "got their act together" using a centralized and directed model, very different from Cambridge but appropriate to their own circumstances. Other British universities generally did not, because they did not clearly define their objectives and policies, or because they implemented their approach badly, for instance by trying to get directly involved in commercialization.[33]

Another study of British science parks, made at about the same time, focuses on the firms. It concluded that science parks had been of no particular assistance to non-academic firms, which would have done just as well elsewhere. On the other hand, they had positively helped academic start-up companies, which might not have come into being without them; but these companies had achieved poorer job-creation rates than other firms. These academic firms seem to need specific kinds of help, including small-scale venture funds, top-quality proactive advice, practical assistance on exports, joint ventures and subsidiary development, and deliberate interface between university research and industry to facilitate transfer. These might represent a good investment of public money, the study suggests, because high-tech firms show a better than average survival rate.[34]

So was Cambridge *sui generis*? To some important degree, undoubtedly so. The particular combination of circumstances could probably have been replicated nowhere else in Britain, or even in the world. But there were lessons that could be learned by many other, less established and less prestigious, university cities. Other places followed subtly different models and made them work, with less spectacular but still respectable results.

Cambridge seems to offer at least three lessons. One is the need to build up a network of individuals and institutions – a university, and in particular certain parts of it, a city council, a bank – that interact in certain positive ways. The recipe for that cannot simply be replicated from Cambridge or any other successful place; it has to be found, by a process that involves a great deal of serendipity, in each successive case. It may however be easier to find in a certain scale and type of community, such as Cambridge, than in other places. In particular, it does underline the need to start in a place which (to paraphrase Gertrude Stein) has some There There. There must be a structure of existing social institutions, capable of developing a synergy, in the first place.

Second, the process takes time. Though Cambridge seized on the opportunity around the year 1970, the seeds of the process had undoubtedly been germinating for some years previously. And the main impacts did not come for a decade or so. Further, they may still be happening. The full impact of the Cambridge phenomenon, as of other similar phenomena analyzed in this book, may not be known for years or decades. All we can say is that some of the impact was already significant in about 10 to 15 years from the start.

The third point to underline about Cambridge is that government did not seem to matter much at all. The phenomenon occurred without apparent state intervention, save perhaps in the very indirect sense that state agencies (for instance, the British Ministry of Defence) may have been clients for some of the intellectual products. Certainly, there was no sense in which systematic government procurement helped build the area, as earlier it had helped create London's Western Crescent. And Cambridge, just as much as the Crescent, was in direct contradiction to British government's regional development policies. All that happened is that – especially after 1980, when the Thatcher Government effectively abandoned such policies altogether – the State simply stood back and watched the phenomenon happen.

HSINCHU: TAIWAN'S SCIENCE-BASED INDUSTRIAL PARK

Hsinchu Science-based Industrial Park is located on the West Coast of Taiwan, 40 miles (70 kilometers) south of Taipei, and 4 miles (6 kilometers) west of the city of Hsinchu, one of the least industrialized areas in the developed, western section of the island (Figure 5.3). The Park was planned and established by the national Government, and started operation in 1980. It extends over 5,189 acres (2,100 hectares), and includes industrial, residential, and research zones, as well as public utilities and land reserves for future expansion, in a clean, pleasant environment typical of a former agricultural area.

The Government's goal was to attract 150 to 200 high-technology firms and to create between 30,000 and 40,000 jobs by the year 2000. When we studied the Park in January 1989, there were about 17,000 employees working in 70 firms, most of them Taiwanese small and medium electronic companies, although several medium-sized American firms, mainly of ethnic Chinese origin, were also located in the Park. There are also two national universities, the National Chiaotung University and the National Tsinghua University, and a major government research institute, the Industrial Technology Research Industry, whose critical role in the Park's development we will analyze below.

Although relatively isolated from Taipei, the Park's easy transportation connection by highway and railroad makes it possible for many engineers and executives to commute from the Taipei area where they seem to prefer to live because of the quality of its schools and urban amenities. The Park Administration however provides good-quality housing and good neighborhood services at affordable prices to employees of firms locating here. There is little relationship with the old town of Hsinchu, which seems to have been entirely bypassed by the new development. Indeed, the Park was not conceived as an instrument of local economic development at all, but as a national Government demonstration project to foster the so-called "co-operation triangle" between government research institutes, universities, and private high-technology firms, under the auspices of the Ministry of Economy.

A Government-led technological development project

Hsinchu was thus entirely a creation of Taiwan's national Government, which invested heavily in the development of the Park and still manages its

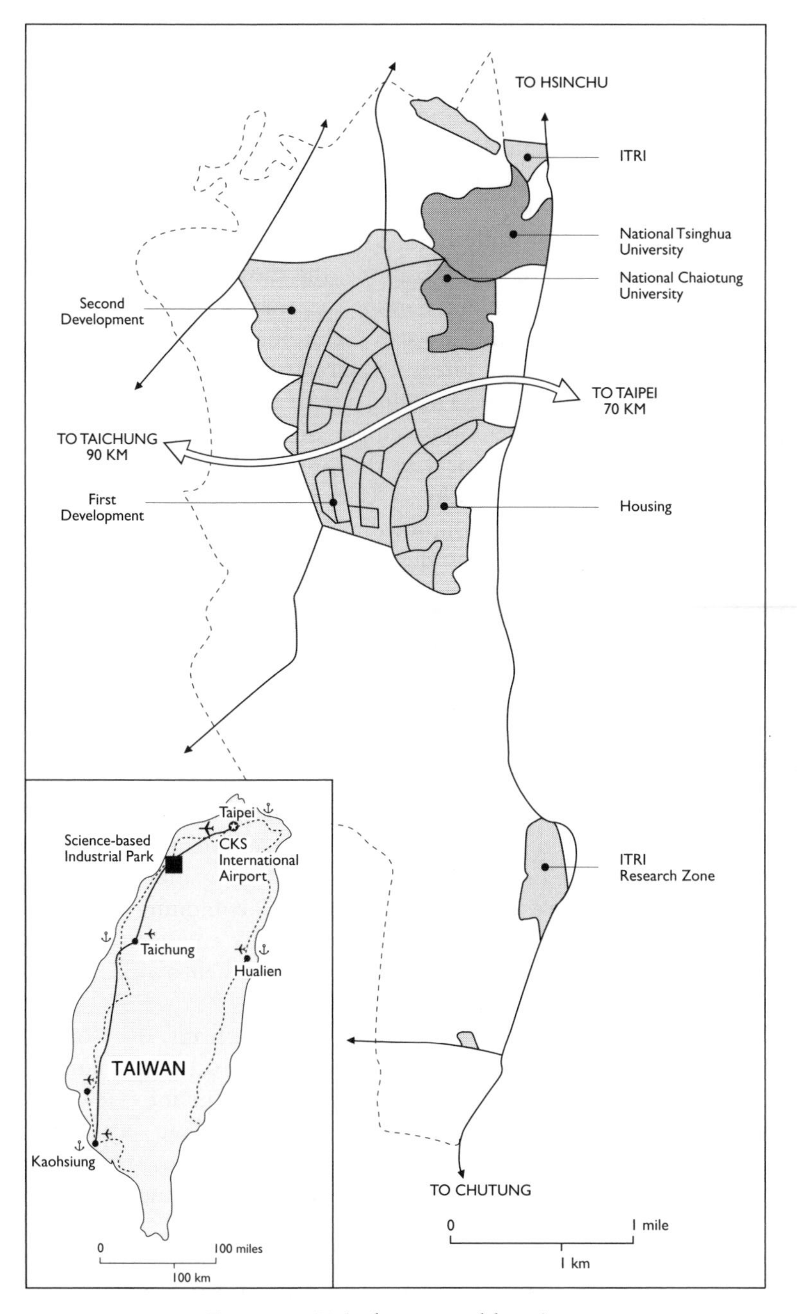

Figure 5.3 Hsinchu: general location
Source: Hsinchu Research Park

daily operations. The Government procured land, built roads and facilities on the site, and developed housing and residential services to house the personnel of the firms locating there. The initial investment in the period 1982–5 was $200 million (US). The data we obtained for 1988 showed an annual expenditure of $440 million (US) in infrastructure.

A major factor behind the location of the Park in Hsinchu seemed to have been the presence here of the Industrial Technology Research Institute (ITRI), a major technology transfer institution created in 1973 by the Ministry of Economy and fully funded by the National Science Committee. ITRI includes five research organizations, the most prominent of which is ERSO (Electronics Research and Service Organization), an institution that has played a decisive role in Taiwan's technological development in electronics. In 1973 ITRI bought integrated circuit design technology from the American multinational RCA. The aim was to transfer design capability to Taiwanese companies; a senior executive of RCA was recruited by the Taiwanese Government to supervise the project. Forty young Taiwanese engineers were sent for 18 months to RCA centers in the United States for advanced training. The first group of these engineers came back to Taiwan in 1974, and that very year they designed the first integrated circuit to be entirely made in Taiwan.

On the basis of this group of engineers, ITRI formally founded ERSO in 1976 and located the new institution in Hsinchu, to provide the researchers with a better environment. ERSO started manufacturing chips for watches, later upgrading production to customized chips for calculators and telephones. In the late 1980s, ERSO was manufacturing chips for microcomputers and peripherals, consumer electronics, and telecommunications, including the design of memory chips such as high-speed RAMs, and MOS RAMs. ERSO was used deliberately by the Government to diffuse microelectronics technology among Taiwanese firms, by organizing seminars and training programs where ERSO's technology was communicated to private firms. In addition, ERSO encouraged some of its engineers to leave the Institute after some years of training and to set up their own companies, in some cases locating in the new Industrial Park.

To ensure the linkage between ERSO and private firms, the Government funds only 50 percent of the Institute's budget. The other half comes from service contracts and technology sales to firms. In our interview with the director of the VLSI division of ERSO he did not see any conflict of interests, because spin-offs from ERSO to the private sector had the effect of upgrading the technological level of Taiwan's electronics industry as a whole.

In addition to providing the basic research and training infrastructure on which Hsinchu Park is based, the Government also devised a number of policies to attract firms into the Park. Firms in Hsinchu enjoy greater fiscal incentives than companies in the export processing zones. These include a five-year tax holiday, a maximum income tax rate of 22 percent, duty-free imports of machinery, equipment, raw material and semi-finished products, and capitalization of investors' patents and know-how as equity shares. In addition, these benefits are not constrained by regulations that normally forbid tax-exempt manufacturers to sell in the domestic market. Tax exemptions, only available in Hsinchu, also include income generated from patents, logo rights, and consulting services. The Government also provides low-interest loans for the purchase of machinery, and makes possible joint-venture options up to 49 percent of total paid-up capital.

Where necessary, the Government also directly enters into industrial production, establishing joint venture companies with private capital. Indeed, the only three Taiwanese companies producing integrated circuits all have Government participation, and all three are located in Hsinchu: besides ERSO, they are United Microelectronics Corporation (32 percent government owned) and TSMC (47 percent government owned). In fact both were developed by engineers originally trained at ERSO.

The Government also offers on-the-job training programs for young engineers belonging to the firms located in the Park, through grants from the National Science Council to undergo training at ITRI or at the two universities located in Hsinchu. Last but not least, the Park's public administration provides the firms with a variety of services, including factory buildings at low rentals, housing, a health clinic, a recreational complex, a shopping center, and bilingual (English/Chinese) schools.

Hsinchu: the overseas connection

The only systematic survey of firms located in Hsinchu was conducted in 1985. It showed that domestic capital firms contributed 52 percent of the capital invested in the Park; foreign firms, 44 percent; and overseas Chinese firms, 4 percent. Table 5.4 provides information on the characteristics of the 51 firms located in Hsinchu at that point. There was significant technology transfer from foreign companies attracted to Hsinchu. But a closer look at these companies shows that most of them have Taiwanese Government participation in their capital, and were owned or managed by Chinese Americans. Thus, besides the 4 percent of investment catalogued as originating from overseas Chinese, many of the foreign firms seemed actually to be a part of the recruitment network of the Taiwanese Government in America.

Indeed, it seems that the main reason for the development of Hsinchu was to provide a pleasant, protected environment to attract these companies to Taiwan. To develop a reverse brain drain is in fact a fundamental element of Taiwan's technology transfer policy. A case in point is the director of Hsinchu Park himself, a Chinese American who was with Honeywell for 27 years and was vice-president of Honeywell's R&D Center in 1981 when he was recruited by the Taiwanese Government. The president of the

Table 5.4 Hsinchu: structure of companies in the Science Park

Characters/types of capital composition and resources	*Government capital*	*Government and multinational capital*	*Government, local private, and multinational capital*	*Government and local private capital*	*Local private capital*	*Multinational capital*	*Local private and multinational capital*
Companies categorized by capital resources and composition	1–18 (18)	19–24 (6)	25–28 (4)	29–34 (6)	35–47 (13)	49 (1)	50 (1)
Transferring technologies from the ITRI, or developed in cooperation with the ITRI	18 (1)	21,24 (2)	27,28 (2)	29,31,34 (3)	35–38, 42,44,45,48 (3)	(0)	(0)
Transferring technologies from the Center for Biotechnology Development	1–10 12–17 (16)	19,20,21, 22,23 (5)	(0)	30 (1)	(0)	(0)	(0)
R&D occurs abroad, or technologies are transferred from MNCs	1–10 12–17 (16)	19,20,21, 22,23 (5)	25,28 (2)	29,30,34 (4)	40 (1)	(0)	50 (1)
R&D occurs in Taipei	(0)	(0)	(0)	30 (1)	35–38,45 (4)	49 (1)	(0)
Product development occurs in the Park	8,9,11,18 (4)	20,21,23 (3)	26,27 (2)	29,32,33 (3)	40–43,47 (5)	(0)	(0)
Focus on the local market	10,11,13, 14,15,16 (6)	19,22 (2)	25,26 (2)	30,32,34 (3)	40–42, 46,48 (6)	(0)	(0)
Focus on the international market (including OEM or subcontracting for mother countries)	1,2,3,5, 6,8,9 (7)	20,21,23 (3)	27,28 (2)	29 (1)	36–38, 43 (4)	49 (1)	50 (1)

Source: Kwan-fu Chen (1986) *High-tech Industry Development and the Spatial Division in Taiwan: A Case Study on the Hsinchu Science Park, Taiwan*: 115, Taipei: National University, Department of Urban Planning.

Note: There are 50 companies listed in the table, which are here given a number (from 1 to 50) rather than listing the names, thereby making it easier to see the overlapping characteristics of the companies and the logic the author used in the table

Institute for Information Industry, one of the ITRI organizations, had spent two years at Fairchild and another seven years at IBM as a senior scientist. Through the networks established by these and other persons in the Chinese-American community of top high-technology researchers, it appeared that in 1989 about one-quarter of the 70 firms located in Hsinchu were a direct expression of a Chinese-American interest in linking up with the booming high-technology sector in Taiwan.

A concentrated, highly-skilled labor force

At the time of our study, in 1989, there were about 17,000 workers in Hsinchu Industrial Park, 50 percent of them engineers, and another 50 percent assembly-line operators; a much higher occupational profile than in other high-technology areas in the newly-industrializing countries. Ninety percent of employees had at least a high school diploma, compared to 40 percent for Taiwan as whole. Thirty percent of the workers, according to official statistics, were engaged in R&D activities. Of all workers, 47.8 percent were college graduates working as technicians and administrators, and an additional 10.6 percent held Masters or Ph.D. degrees. ITRI alone had 4,000 engineers, including 850 engineers at ERSO, the electronics institute, of which 40 percent had Masters or Ph.D.s. Engineers in the institutes and firms were mainly recruited from Taipei, while the majority of assembly workers were recruited from Hsinchu's surrounding areas.

Why Hsinchu?

In our interviews with company executives, the reasons given for originally locating in Hsinchu were unconvincing. They seemed, in our opinion, to come down to one fundamental reason: to please the Government. But there was an added twist: once in existence, Hsinchu did become a "high-tech address" in Taiwan. Easy access to the harbor (50 miles, 80 kilometers away) and the international airport (25 miles, 40 kilometers distant) and availability of land did not appear to be overriding reasons, since many other sites in the periphery of Taipei could satisfy these requirements. But given the importance of a good connection to the Government in doing business in Taiwan, as well as the special package of incentives provided in Hsinchu, foreign firms did not have any major objection to forming a part of a governmental technological development scheme.

Yet, once in Hsinchu, a more important locational factor emerged as critical in maintaining a presence there: Hsinchu itself. After a few years, most firms developed a strong network of suppliers and clients, as well as easy recruitment of highly-skilled engineers. In the words of Taiwan's director of the multinational company Philips: Taiwan's advantage in attracting foreign investment "has shifted from cheap labor to cheap brains."

Hsinchu seems to have succeeded in providing simultaneously an adequate labor supply, a network of ancillary companies, and a potential market. In short, Hsinchu in 1989 was on its way to becoming a high-technology industrial district. Linkages between firms, universities, and research institutions seem to have been critical for the success of a project that succeeded, against all the odds, in establishing an advanced production site in an underdeveloped area of an underdeveloped island off the coast of China.

Linkages

Firms in Hsinchu entertain close contacts with the two major universities in the area, although their recruitment extends beyond Hsinchu into the Taipei area. There are also frequent research contracts awarded to the universities by the Park's firms: an average of $180,000 (US) goes to Tsinghua's budget every year. There are a number of other programs involving the two universities and the companies in the Park. For instance, the companies organize workshops to inform university students about recent developments in high-tech industry and provide the universities with advanced equipment and experience in industrial production; the universities hold open house sessions to inform the companies of their research achievements; the universities also provide training programs for the employees in the Park. Again, although the locational factor has made it easier for the companies in the Park and the neighboring universities to establish these connections, there is also no major problem for other universities to create such connections with the companies in the Park; or for the two universities to establish such connections with companies outside the Park.

Nevertheless, it is still debatable whether such cooperation between the universities and the companies in the Park plays any substantial role in the connection between research and its application, or if the spatial proximity between the two is as crucial as it might be in other cases. The Dean of Academic Affairs of Tsinghua University suggested that there is no significant advantage for the University and the Park to be in the same area, because Taiwan's industrial production has not involved any "real" R&D activity; so the companies do not need help from the University. He argued that the companies in the Park are engaged in manufacturing high-tech products rather than in high-tech research and development. Even if companies in the Park need certain kinds of science and technology information or assistance, ITRI will do most of the job for them.

As earlier noted, the Government-supported technological and scientific research institutes have been indispensable in Taiwan's high-technology industrial development. In many cases they have played a more important role in providing technological expertise to companies than have the universities. The major characteristic of these research institutes is their emphasis on commercially viable applications of research, that will prove internationally competitive and thus exportable. The director of the Industrial Technology Research Institute considered himself a businessman rather than a researcher or a research manager. The research groups of ITRI maintain a close relationship with clients as a check on commercial viability. Many research projects originate in consultations with clients. When a specific project is decided upon, ITRI may invite firms other than the original proposer to participate. Regardless of the number of participants, they all take part in projects, share the costs and obtain full use of the results.

In 1984, for instance, the Electronics Research and Service Organization (ERSO) added five new computer projects to its program. These projects, estimated to cost $5 million, included a 32–bit personal computer using MS-DOS and Unix operation systems, a color printer, a portable Chinese typewriter with a 5,000-character storage capacity, a 5.25-inch 50-Mbyte hard disk drive, and a 3.5-inch 20-Mbyte hard disk drive. One project included five participating firms, the others only one or two. The ERSO procedure gives firms several options for meeting R&D requirements: they can go it alone, they can seek help from Government-supported research institutes, or they can pool resources with others to produce technology that might be shared with competitors. ERSO's research tackles business as well as technical questions, such as quality control, marketing strategy and yields. ERSO also gives production and marketing guidance after finishing the research and turning over the licensing rights to its clients. Further, ERSO also produces IC components for companies.

As to the issue of spatial proximity between ITRI and the companies, again we found that it is a necessary but not sufficient element for Hsinchu's success. ITRI has provided consulting services and IC components to companies both inside and outside the park. At the same time, ITRI's services do not exhaust the need for technological aid on the part of the companies in the park. Given the small size of the island of Taiwan (13,892 square miles, 35,980 square kilometers) and the well-developed communication networks, connections with research institutes and companies outside the Park are both possible and necessary for institutes and companies alike. This situation also applies to the relationship between ITRI and the universities. ITRI has been attractive to university graduates because, unlike private companies, ITRI does not require work experience in its recruiting policies. Every year ITRI recruits domestic engineering students from all the major universities in Taiwan. The graduates of the two universities near the Park do not outnumber the graduates from other universities in Taiwan.

However, this does not suggest that there is no need for spatial proximity for these actors in the process of high-technology industrial development in Taiwan. Given the characteristics of an industry of which information is the most important raw material and product, there is a great need to be able to exchange personal views on some start-up's breakthrough last week, or on the new project initiated by some company. A research manager pointed out that the biggest locational advantage of the spatial concentration of the companies and the research institutes is that the researchers and engineers meet each other constantly in a restaurant located in the Park. The restaurant is one of the very few fashionable restaurants in Hsinchu, and it has become the place to go for the professionals and intellectuals in town. They meet each other at the restaurant, and often engage in spontaneous conversations. Such casual gatherings have provided them with opportunities to exchange ideas and information in a rather informal but very effective way.

In short, the spatial concentration of ITRI, the two universities, and the Science Park in Hsinchu is important, not in the sense of physical coexistence, but through the linkages and social relationships established among them.

Hsinchu Park in the Hsinchu region

In spite of the locational advantages mentioned before, there have been debates as to whether the Hsinchu area is an ideal location for the Park. By 1985, 80 percent of Taiwan's information-related industries were still located in metropolitan Taipei. Many companies locate their production and part of their research activities in the Park in order to be qualified for the special tax exemptions, while keeping their headquarters in Taipei as centers of research, decision making and marketing. Therefore, Taipei, rather than Hsinchu Science Park, seems to be the real center of high-tech industry in Taiwan. Located 40 miles (70 kilometers) from this center, the Park is inevitably handicapped by a certain deficiency of information flow.

Another concern about the location is its lack of urban culture and activities. Compared with the congested city of Taipei, Hsinchu is greener and more spacious. However, as one of the least industrialized cities of Taiwan, Hsinchu has long suffered from out-migration and lack of industrial and commercial investment. The city, 4 miles (6 kilometers) away, still keeps the pace and atmosphere of a small town rather than that of an industrial city. The Park and the universities have tried to provide the daily needs of the students, faculties, and engineers, with restaurants, theaters, bookstores, a gymnasium, and other facilities; and a limited array of services and small shops has grown around the institutes and their residential areas.

Further, if engineers and university faculty members want to do serious shopping or to attend cultural activities not available locally, it is easy for them to get on the highway to Taipei. Many of them have their families living in Taipei and they either commute daily[35] or spend weekends in Taipei. By and large the technology buildup has therefore been isolated from the city in which it is located, and there is no significant sign of major connections between the technology complex and local economic development.

The high-tech technology complex of HSIP is also isolated from the city administratively. HSIP's administration is subordinated to the National Science Council. Budgeting for the Park is directly related to the National Science Foundation, bypassing the Hsinchu city and county government. This administrative separation between Park and locality has generated some conflicts, notably over taxation. The companies in the Park pay a limited amount of tax to the local government, but, because most companies keep their headquarters and sales in Taipei, the major portion of the corporate income tax goes there. Meanwhile, as the industries in the park keep growing, the Park administration plans to expand the Park. Since the local government is unlikely to benefit from this growth (not to mention the loss of control of valuable land), it has not expressed any major interest in cooperating with the Park administration. This has further enhanced the isolation of the high-technology complex from the rest of the region. As Taiwan moves toward democracy, and as local authorities play an increasingly active role, potential conflicts between local and national governments over the control of the Park's development and benefits could sour the future business atmosphere.

Hsinchu and the developmental State

Hsinchu epitomizes Taiwan's drive to develop high-technology manufacturing. In spite of its limited importance in the broader context of the island's industrial development, the building of an industrial complex on the basis of a government program, and of government-established research institutions, shows the direct connection between development policies and the new industrialization process. By developing forward and backward linkages in the Park, private firms have contributed decisively to economic growth and technological upgrading. While much of what happened in Hsinchu could have happened in Taipei, the fact is that a government-initiated project became an entrepreneurial industrial center, successfully supported by some of the most advanced technological centers in Taiwan. It was on the basis of such development-oriented policies that Taiwan modernized and upgraded its industrial structure to become an aggressive competitor in high-technology trade in international markets. In spite of its largely bureaucratic origins, Hsinchu provides material evidence of the impact of the developmental State on the new shores of the world economy.

CONCLUSION

The analyses presented in this chapter show that it is indeed possible to create new industrial spaces based on high-technology firms, even in relatively remote locations, such as North-Central Taiwan, or in previously underindustrialized regions, such as the Côte d'Azur or Cambridgeshire, given the presence of certain critical factors and the support of local, regional, or national authorities. Among the critical factors can be listed the presence of research and training institutions, favorable tax and credit incentives, availability of industrial land, a local labor market with quality engineers and technicians, a good transportation system, and adequate telecommunications. Environmental quality, bureaucratic flexibility, and a good locational image also enhance the attractiveness of a technology park.

Quite different, however, are the conditions for the "success" of a technology park. Success needs to be defined specifically in relation to the objectives pursued in the development of each park. Most difficult of all is the creation of linkages and synergistic interaction between the park's various components. In general, industrial high-technology parks are made up of three components: public research centers and universities; large firms; and small and medium firms. Experience seems to show that the relationships between large firms on the one hand, and public and university research centers on the other, prove most difficult. They tend, as in Sophia-Antipolis, to generate two different cultures, often secluded and sometimes hostile to each other.

Thus, the critical question for the generation of synergy proves to be the relationship between small and medium firms and each one of the two other components. When large firms link up with a network of small firms, as in the case of Sophia-Antipolis's information technology complex, a new productive dynamism can appear. Such is also the case in the relationship between Cambridge University and a number of spin-off private firms in that city, and between ERSO's Electronics Laboratory and Taiwanese firms in Hsinchu. On the other hand, the isolation of the pharmacology–biology–chemistry complex in Sophia-Antipolis or the relative self-sufficiency of foreign electronic companies in Hsinchu provide evidence that the mere fact of location in a technology park does not generate synergy by itself. The differences are due to the diversity of firms' strategies, as well as to the stage of development of the park at the moment of the location of the firm.

Another crucial factor is the role of government in the development of a park. Without such support, it seems very unlikely that technology parks could ever develop in previously underindustrialized regions. In Hsinchu the Government was of course primarily responsible for the Park, as is generally the case in the Asian Pacific Region, the supposed epitome of laissez-faire economies. In Sophia-Antipolis, the initial impulse came from a public Grande Ecole, with the support of the local authorities, and later received help from the national Government and from nationalized companies, such as Air France, which were lured into the venture by the Government. Even in Cambridge, a semi-spontaneous development spun-off from the University, industrial development was blocked for some time by obstructive local planning policies, and could only take off after the revision of the Cambridgeshire Development Plan in 1971 to follow the recommendations of a 1969 University report which advised the establishment of a "Science Park" in the area.

Thus, the role of government and/or a university seem to be crucial for the establishment and growth of a park. Yet the ability of the park to generate new economic dynamism depends also on the awareness of governments that they must let parks grow primarily through private investment and entrepreneurial initiative. The greater the role of private firms in the development of the park, as in Cambridge, the greater the chances of generating growth and innovation. At any rate, in the three cases we have studied, and in many more experiences we are aware of around the world, technology parks that meet these basic conditions do create jobs and attract inward private investment, thus playing an important role in industrializing or reindustrializing regions.

But judged by the criterion of creation of an innovative milieu, in the strongest sense of that concept, few technology parks are a success. Indeed, even the three experiences analyzed here are only partial successes on this count. Economic and technological synergy requires the development of industrial linkages and information networks that took place only on a limited scale among small firms in Cambridge, in the information technology complex in Sophia-Antipolis, and between the Government Laboratory and small and medium firms in Taiwan. But in all three cases, the existence of a technology park in its various forms triggered a process of industrial growth and technological upgrading of the local economy, literally putting these areas on the map of the new industrial geography.

6

JAPAN'S TECHNOPOLIS PROGRAM

The technopolis program is distinctive, because – unlike any other project considered in this book, indeed any such project anywhere – it is so vast. It is a national plan, master-minded by MITI – Japan's Ministry of International Trade and Industry – to create an entire series of new science cities in the country's peripheral areas, in order simultaneously to promote new technologies and develop lagging regions. The aim is thus audacious: it is literally to create a whole set of new innovative milieux, thus transferring the power to generate new commercial technologies from the heart of the country, to places that previously lacked that capacity. The plan seeks to achieve this aim though multiple strategies: concentrating public and private research institutes, promoting hybrid technologies, upgrading local university laboratories, establishing technology centers, funding joint R&D projects, and providing R&D funding.[1]

The background and starting-point will be found in Chapter 7: it is an acute regional imbalance, which has worsened since World War Two. Tokyo by the early 1980s had one-quarter of the Japanese population; Japan's three major metropolitan areas – Tokyo, Nagoya and Osaka, strung out along the intensely-urbanized 300-mile Tokaido corridor – had close on half.[2] But the concentration of innovative capacity, which has driven Japan's incredible rise to global economic power since 1950, is even more concentrated: almost 80 percent of all corporate labs, 70 percent of all scientists and 60 percent of all university professors were in the Greater Tokyo and Osaka areas. Greater Tokyo alone had about half of total manufacturing by shipment, and an even higher concentration of new high-tech factories; it had more than 50 percent of company R&D;[3] it accounted for 65 percent of all computer installation and 61 percent of information-processing employment in 1984.[4] And during the 1980s, as major Japanese corporations have become truly transnational, this concentration has if anything accentuated.[5]

MITI AND THE JAPANESE DEVELOPMENTAL STATE

To understand the technopolis program fully, the essential starting-point is that Japan is not, and never has been, an ordinary market capitalist state. As several excellent studies have shown, its economic system can more accurately be described by a term that Japanese economists use: it is state monopoly capitalism, planned and directed by bureaucrats in an extraordinarily close association with the big conglomerate corporations that dominate the Japanese economy. The over-worked phrase Japan, Inc. describes it; but better is Chalmers Johnson's term, the developmental state. It is a model – later followed (with differences) by Taiwan, the Republic of Korea and Singapore – of a new kind of planned economy. Ironically, it has been as conspicuously successful as the better-known kind – the Stalinist command model – has been a failure.

Its origins go a long way back, farther than most people think: to Japan's opening-up to the world, in the Meiji restoration of 1868. And perhaps even farther: the bureaucrats, who from the start administered the modernization process that followed, could effectively trace their lineage back to the *samurai* class of old. The most important part of this bureaucracy, which was created as the Ministry of Commerce and Industry in 1925, became the Ministry of Munitions in 1938 and controlled Japan's war machine; it was reborn as the Ministry of International Trade and Industry during the postwar American occupation. At first, interestingly, it tried to control the economy in almost Stalinist fashion; but, after a series of bruising battles with business, it reached the remarkable system of planning by consensus that has guided the Japanese economic miracle.

It has been able to do this because of some unique features of Japanese life. In the first place, the MITI bureaucracy has remarkably close ties with other key Japanese institutions, especially with Tokyo University, from which most of its members (and virtually all of those who achieve top positions) are recruited, and also with the management of the major corporations, who recruit retired bureaucrats via the custom of *amakudari* ("descent from heaven").[6] And secondly, MITI has been able to control its developmental state through an iron grip on the supply of industrial capital: the great Japanese conglomerate companies *(keiretsu)* are essentially financed not through the stock market, but through loans from their group bank, which in turn are financed from Japan's Central Bank through a pattern of systematic over-lending; on top of that, capital is available from a large pool, the Fiscal Investment and Loan Plan (FILP), derived largely from tax-free Post Office savings accounts. MITI has built upon this control a system of extra-legal "administrative guidance" to industry, which it developed to overcome the loss of its direct controls during the trade liberalization of the early 1960s.[7]

Since 1949, in fact, MITI has several times shifted its policies to accord with Japan's evolving economic position as well as with both internal and foreign political pressures. During the early 1950s, the era of postwar reconstruction, it skilfully manipulated the 1949 Foreign Exchange and Foreign Trade Control Law – which General MacArthur had promulgated in the naive belief that it would be temporary – to protect struggling industries by high tariffs, quotas, import quotas, strict inspection procedures and investment controls, plus selective financial assistance to industries targeted for growth. In the high-growth era of the 1960s, working closely with big corporations, it systematically nurtured new industries, *ikusei*.[8] That era came to an abrupt end with a wave of political discontent occasioned by concerns about overcrowding and environmental pollution, closely followed by the two oil shocks of the mid-1970s, which together produced a severe blow to MITI's image and prestige. But MITI bounced back by embracing the ideas developed in 1969 by Naohiro Amaya, a young bureaucrat: it would now encourage the development of new high-technology industries based on high value added, and simultaneously it would promote the encouragement of out-movement from Tokyo and Osaka. In 1979 the 1949 law was amended, greatly liberalizing trade.[9]

From that point on, MITI was no longer in the business of catching up with the world industrial leaders; henceforth, it would play the infinitely more difficult game of becoming, and remaining, a pioneering world leader.[10] So, since 1980, MITI has systematically moved to "innovate at the frontiers." It has targeted 14 industries: aircraft, space, optoelectronics, biotechnology, computers, robotics, medical electronics, semiconductors, word processors, new alloys, fine ceramics, medicine, software, and electronic machinery. It works to nurture these infant industries until they are strong enough, then lets the market take over. More than 30 parallel-track R&D projects, under several different agencies – including MITI's Agency of Industrial Science and Technology (AITI), which runs 16 labs with 3,500 researchers – develop subsidized projects, large-scale projects too risky for individual companies, and National Research Projects which gather together talent from leading companies and government laboratories.[11]

TECHNOPOLIS AND REGIONAL DEVELOPMENT

The drive to achieve systematic leadership in world technological innovation is thus one key thrust behind the technopolis project. But there is another, equally important, aim: regional development, to reduce the geographical imbalance, and in particular the dominance of Tokyo. Regional policies are nothing new in Japan: in 1962, MITI's Comprehensive National Development Plan sought to divert population and industry from the three large metropolitan areas and to decentralize government, education, and industry, with the aid of 15 new cities serving as regional growth poles; in 1969, the New Comprehensive National Development Plan of 1969 aimed to build a national network of expressways and Shinkansen (bullet trains), plus large industrial projects; in the mid-1970s, Prime Minister Kakuei Tanaka aggressively implemented this, with plans to link Japan by a network of bullet trains, highways, telecommunications networks and new information cities;

in 1977, yet a Third Comprehensive National Development Plan aimed to improve the quality of life in certain designated areas. But all foundered through high costs and financial crises, and certainly none seemed to work. The present policy is composed of three tiers: namely, Tsukuba Science City (considered in Chapter 4); a set of dispersed "research cores"; and the technopolis program.[12]

The origin of the technopolis program lay in late 1979, when MITI began studying the possibility of creating a Silicon Valley in Japan. Officials had been impressed by Dr Edwin Zschau of the American Electronic Association, who had spoken of "the process of innovation." They brought together Professor Takemochi Ishii of Tokyo University Engineering Department, Hajime Karatsu of Matsuhita Communications Company and Tatsuo Takahashi of MITI's Industrial Location Division, to analyze the reasons for Silicon Valley's success. They identified a familiar list – research universities, industrial parks, a pool of engineering skills, venture capital, investment banking, management consulting, support services, informal networks – and sought to combine this with Japanese success in existing high-tech concentrations. Neither Tokyo nor Tsukuba Science City appeared to offer a model: the first because it was too competitive and imitative, the second because its researchers were too academic. They heard the word *technopolis* from Toshiyuki Chikami, mayor of Kurume on Kyushu; somewhat ironically, he had borrowed the title of a pop song about Tokyo, which became the number one hit of 1980 in Japan.[13]

The policy was fleshed out in *MITI's Vision for the 1980s*: essentially, an indicative economic plan, produced by the advisory Industrial Structure Council – comprising representatives of academic and industrial circles, consumers, and labor unions – in 1980. It described the vision:

> "Technopolis" (technology-intensive city) is a city that effectively combines an industrial sector composed of electronics, machinery and other most advanced technologies with an academic and a residential sector. This concept aims at promoting regional development and creating a new regional culture under the lead of industrial and academic progress. A possible model scheme in and after the 1980s, it differs in its basic approach from the conventional ideas of regional development centering on land utilization and infrastructural improvements.[14]

Combining elements of Silicon Valley and Tsukuba, and even much older garden city notions that the Japanese had imported from England early in the present century, the vision featured research universities, science centers (technocenters), industrial research parks, joint R&D consortia, venture capital foundations, office complexes, international convention centers, and residential new towns.[15] Unlike earlier Japanese regional development exercises, it rejected public works projects in favor of a "soft" infrastructure of trained people, new technologies, information services, venture capital, and telecommunications services.[16]

For MITI, however, there was another objective of the program: it was to discourage offshoring, also known as hollowing-out, of the Japanese economy. Cabinets in the mid- and late 1980s stated their intention to use their power to stimulate a more inward-looking development process: instead of exporting manufacturing plants to Taiwan or Malaysia, Japanese companies would be encouraged to move them to the underdeveloped peripheral regions of Japan.[17] Coupled with this was the notion of a rural social development plan including production, education and quality of life, aimed at developing a new rural culture that would cause migrants to come back from the congested metropolis to their regions of birth.[18]

FROM CONCEPT TO ACTION PROGRAM

A response was not long in forthcoming: the Laws for Accelerating a Technopolis Based on High-Technology Industries (Technopolis Laws) passed the National Diet in 1983, and the project started under a Technopolis Committee, with a tight timetable: to choose the sites by 1984, to complete construction of the physical infrastructure by 1990, and to complete development of each technopolis by the year 2000, with the aim of generating a "Techno-Archipelago" in the twenty-first century. The sites to be chosen had to meet certain rigorous criteria:

1 a total area (on one or more sites) of 1,300 square kilometers (500 square miles) or less;
2 existing enterprises with potential for high-tech development;
3 easily available industrial sites, available water and residential areas;
4 an existing city (Mother City) with 150,000 or more people;
5 an existing university with high-tech education or research;
6 access to high-speed transportation facilities giving a one-day return trip from Tokyo, Nagoya or Osaka.[19]

It was important that the technopoles would be not merely high-technology production centers, but would also develop local innovative R&D capacity to help trigger the development of such industries locally. No one, not even MITI, knew precisely how to do this: the policy was simply to "target" particular high-technology industries from MITI's general priority list – aircraft, space, optics, biotechnology, medical electronics, robots, integrated circuits, computers, software, data processing, fine ceramics, medicine and medical supplies, and industrial machinery – for development through an initial pump-priming subsidy to stimulate R&D.[20] The R&D capacity would be built up in two parallel ways: first through relocating existing high-technology industries from the congested metropolitan areas, and second through assisted self-development of existing local industries. Thus the program was supposed to benefit not only big enterprises, which would be more likely to relocate or set up branch plants, but also small and medium ones, so assisting "grass roots technological revolution" in the technopolis areas.[21] But this latter would also have a dual parallel-track character: early on, the emphasis would be on transferring R&D capacity, especially electronics and mechatronics, into existing industries; later – for this would take more time – the aim would be to develop frontier or creative R&D that might eventually create new industries.[22] Basic private-sector research, the argument went, could be relocated outside the big metropolitan centers because its locational needs – access to scientific information, access to headquarters functions to enhance coordination, and access to the amenities of metropolitan areas – could be provided in the technopolis Mother City.

There was one further feature: the technopoles were to be chosen by open competition among the prefectures, who would thus be encouraged to provide their own local incentives. This marked a novel approach for MITI: instead of a top-down approach, the prefectures would play the crucial role in planning and constructing the technopoles, with MITI's role limited to setting basic criteria and providing technical assistance, advice, tax incentives and loans from the Japan Development Bank. The idea proved all too successful: the study group had recommended two or three projects, but there was a stampede in which 40 out of 47 prefectures volunteered and 19 were chosen, with another seven selected subsequently to bring the total (in 1990) to 26 (Figure 6.1). Not all of them met the criteria – a fact that caused the press to accuse MITI of caving in to political pressures, both from the Diet and from large high-tech companies, construction groups, and consultancies who stood to benefit.[23] And one crucial element of the entire program has undoubtedly been the subtle tensions between MITI, the pork-barrel politicians of the LDP establishment, and the bottom-up provincial interests – given that the prefectural governors are also major LDP politicians, some of whom have close ties with MITI.

Certainly, though the program was originally intended to pass the main responsibility down to the localities, in practice it has meant fairly generous government spending. MITI subsidizes frontier R&D through the Small and Medium Business Agency, and funds for technology development through the National Academy of Industry and Technology, plus relocation advice. MITI's technopolis budget in 1985 was ¥1,485 million apart from frontier R&D, relocation and other promotional efforts; the Ministry of Construction was responsible for the hard infrastructure.[24] Completion of the program, originally scheduled for 1990, was in fact delayed because of tight finances. Even so, the Ministry of Construction estimated that the construction costs for 11 of the technopolises averaged $200 million each by 1990.

THE TECHNOPOLES IN OPERATION

The location and characteristics of the 26 technopolis sites are summarized in Table 6.1 and Figure 6.1. Though the main emphasis is clearly on the

Figure 6.1 Japan: The technopolis sites
Source: MITI (1990) "Progress in the Technopolis Programme"

peripheral regions, the sites vary greatly in their relationship to the dominant Tokyo–Osaka (Tokaido) axis: while a few, such as Kofu and Hamamatsu, are very close to and almost part of it, the majority are more than 300 kilometers (180 miles) from Tokyo, and 10 are off the main island of Honshu – six on the southern island of Kyushu, two on the small island of Shikoku off western Honshu, and two on the northern island of Hokkaido. In terms of their size and population they are somewhat more homogeneous: areas range from 30,000 to 140,000 hectares (115 to 550 square miles); populations of the main (mother) city range from 175,000 up to 728,000, with most close to the middle of the range; most significantly, the targeted industries conform closely to MITI's general priority list. And in every case, there is a significant research complex at the heart of the site, developed either out of an existing institute, or through creation of an entirely new center.

There is however another important respect in which the technopolis sites are very different.[25] Some, such as Hakodate, can be described as *lagging technopolises*. These lack a substantial manufacturing base secondary industry; traditional local industrial firms lag far behind in introducing new technology.[26] Others, *technopolises successfully facilitating "satellite-type industry,"* like Tohoku and Kyushu, successfully attract outside capital in such industries as semiconductors, new ceramics, general machinery and electronic machinery; but these tend to deprive small traditional firms of skilled workers, causing a crisis for them. The third group, *technopolises successfully transforming "traditional local industry" into "modern local industry,"* like Hamamatsu, have the best prospects for success: in them, "traditional industry" pursues R&D and becomes more technology-intensive. Starting in textiles and lumber processing, Hamamatsu has built a strong base in machinery and musical instruments, and is now developing the automobile industry. According to Noguchi, Nagaoka falls into this group: its firms are successfully introducing high-technology, including robots.[27]

To give a closer picture of the character of the individual technopolis sites, we will describe four of them, which we visited in the summer of 1990. Two are on the main island of Honshu, north of Tokyo – that is, away from the main Tokaido axis but still reasonably close to Tokyo. The other two are on the remote southern island of Kyushu, which however – even before the establishment of six technopolis sites here – had become one of the leading localities for semiconductor production in Japan and in the world.

Sendai

Sendai, 300 kilometers (188 miles) north of Tokyo and the seat of the Miyagi Prefectural Government, is at first an odd site for a technopolis. Only 102 minutes from Tokyo on the Tohuku Shinkansen, it does not appear very peripheral; indeed, local officials argue that with increasing congestion in the Tokaido Corridor, the focus of development is already shifting here. And within the prefecture, primarily in Sendai itself, are 10 universities – including Tohoku University, a research university with a strong reputation in semiconductors – eight junior colleges, and two technical colleges.

The technopolis, approved by MITI in December 1986, includes Sendai City (designated the Mother City, with 800,000 population) and Izumi City, two towns and a village, totalling about 200,000 acres (81,000 hectares or 810

Table 6.1 Designated technopolis areas and features

Area	*Prefecture*	*Mother city*	*Number of cities/towns/ villages in technopolis*	*Population (technopolis area) (mother city)*	*Features of industrial complex*	*Features of R&D concept*
Doo	Hokkaido	Sapporo	3 cities 1 town	285,736 1,542,979	Mechatronics, new materials, biotechnology	Eniwa research business park, etc.
Hakodate	Hokkaido	Hakodate	1 city 3 towns	380,517 320,152	Marine industries, resources utilization industries, frigid area community development	Integrated regional marine research center, urban development center, resource utilization research
Aomori	Aomori	Aomori	4 cities 2 towns 2 villages	604,325 287,597	Mechatronics-biotechnology industries	Local industry research, modern technology research laboratory, institutes of industry and technology, etc.
Kitakamigawa	Iwate	Horioka	4 cities 1 town 1 village	243,601 235,469	Electronics, mechatronics, bioindustry, new materials	Industrial Center (Research Core), industrial technology center, etc.
Akita	Akita	Akita	1 city 2 towns	304,823 284,863	New materials, resource-energy development, electronics, mechatronics-related industries, biotechnology-related industries, etc.	Metal frontier center, local technology center, medical center for the elderly, etc.
Sendai-Hokubu	Miyagi	Sendai	1 city 3 towns 1 village	877,296 700,248	High-tech electric machinery, new materials, bioindustry, city information industry	21st-Century Plaza (Research Core), etc.
Yamagata	Yamagata	Yamagata	5 cities 1 town	502,274 245,158	Mechatronics, biotechnology, fashion, etc.	Yamagata Techno-Create Center (Research Core), Yonezawa High-tech Center, etc.

Table 6.1 continued

Area	*Prefecture*	*Mother city*	*Number of cities/towns/villages in technopolis*	*Population (technopolis area) (mother city)*	*Features of industrial complex*	*Features of R&D concept*
Koriyama	Fukushima	Koriyama	2 cities 3 towns 1 village	403,527 286,451	Microelectronics, new materials technology, biotechnology	Industrial technology center, etc.
Shinanogawa	Niigata	Nagaoka	8 cities 6 towns 1 village	638,509 183,756	High-dimension systems, new materials processing, urban business, fibre industry, fashion	Development Education Research Promotion Center of Nagaoka Technical Univ. (established), Technopolis Development Center, Kashiwazaki Softpark
Utsunomiya	Tochigi	Utsunomiya	2 cities 2 towns	469,944 377,746	Mechatronics, electronics, fine chemicals, new materials, etc.	Mechatronics laboratory, regional industrial institutes, institutes of physics/technology, etc.
Kofu	Yamanashi	Kofu	2 cities 14 towns 5 villages	420,410 202,405	Optoelectronics and systems, highly developed mechatronics, software, new materials	21st-Century Industrial Park (Research Core), industrial technical center, etc.
Asama	Nagano	Nagano	3 cities 6 towns 1 village	333,472 336,967	Highly developed mechatronics, high-quality parts, biotechnology	Ueda Research Park, Sakaki Techno-Center, etc.
Hamamatsu	Shizuoka	Hamamatsu	3 cities 2 towns	619,621 490,824	Photo-industry, musical instruments, sophisticated mechatronics, information communication systems, etc.	Photo-information Technology Integrated Research Center, Electronics Center, life behavior research organs, etc.

Table 6.1 continued

Area	*Prefecture*	*Mother city*	*Number of cities/towns/ villages in technopolis*	*Population (technopolis area) (mother city)*	*Features of industrial complex*	*Features of R&D concept*
Toyama	Toyama	Toyama Takaoka	2 cities 4 towns	568,291 480,110	Biotechnology, mechatronics, new materials, etc.	Toyama Tech. Development Corp., Bioscience Research Center, Modern Technology Interchange Center
Nishi-Harima	Hyogo	Himeji	4 cities 10 towns	716,679 446,256	High-tech industries, medical and social welfare industries, etc.	Life science laboratories, etc.
Kibi Plateau	Okayama	Okayama	3 cities 5 towns	660,183 545,765	Medical, pharmaceutical, and chemical industries, agro-industry, etc.	Enzyme/bacteria bank, experimental organism center, graduate schools, etc.
Hiroshima-Chuo	Hiroshima	Kure	3 cities 2 towns	375,855 234,549	Mechatronics, ship and ocean electronics, home electronics, regional community systems, etc.	R&D organs, modern technology development center, international materials science research center, etc.
Ube	Yamaguchi	Ube	3 cities 4 towns	408,774 168,958	Fine chemicals, biotechnology, new mechatronics, electronics, mechatronics, etc.	Technology promotion corporations, new materials development organs, ocean development organs, etc.
Kagawa	Kagawa	Takamatsu	5 cities 7 towns	635,705 316,661	Ultra-precision measuring instruments, living systems, ocean development, etc.	Experiments/research, life technology experimental cities
Ehime	Ehime	Matsuyama	6 cities 6 towns	813,130 426,658	New materials, fine chemicals, electronics, mechatronics, biotechnology, information	Techno Plaza Ehime, industrial technical information center, etc.

Table 6.1 continued

Area	*Prefecture*	*Mother city*	*Number of cities/towns/villages in technopolis*	*Population (technopolis area) (mother city)*	*Features of industrial complex*	*Features of R&D concept*
Kurume-Tosu	Fukuoka, Saga	Kurume	2 cities 5 towns	332,487 216,974	High-system industry (information-associated industry, community development, mechatronics), new materials, biotechnology, etc.	R&D park, integrated information center, etc.
Kan-Omurawan	Nagasaki	Sasebo	3 cities 3 towns	440,778 251,188	Ocean development-associated instruments, resources and energy development based on mechatronics, etc.	Research Park (Bio wood), laboratories for research in electronics applications for machine technology and for semiconductor applications
Kenhoku-Kunisaki	Oita	Oita, Beppu	4 cities 13 towns 2 villages	281,513 496,963	IC, LSI, new materials, soft engineering, techno-green industry, regional resources utilization, etc.	Regional economy information center, industry-university-government cooperation system, etc.
Kumamoto	Kumamoto	Kumamoto	2 cities 12 towns 2 villages	738,558 525,662	Applied machine industry, biotechnology, computers, information systems, etc.	Research park (Bio wood). Electronics applications machine technology research laboratories, etc.
Miyazaki	Miyazaki	Miyazaki	1 city 6 towns	356,876 264,855	Electronics, mechatronics, new materials, biotechnology (fine chemical, biomass, etc.)	Cooperative Research Development Center, IC laboratory, etc.
Kokubu-Hayato	Kagoshima	Kagoshima	2 cities 12 towns	691,909 505,077	Advanced equipment (electronics, mechatronics), new materials (fine ceramics), regional industry (modern fishing and agroindustry, biotechnology, etc.)	Technology promotion organs, material resources research center, regional industry promotion associations, etc.

Source: MITI, 1990

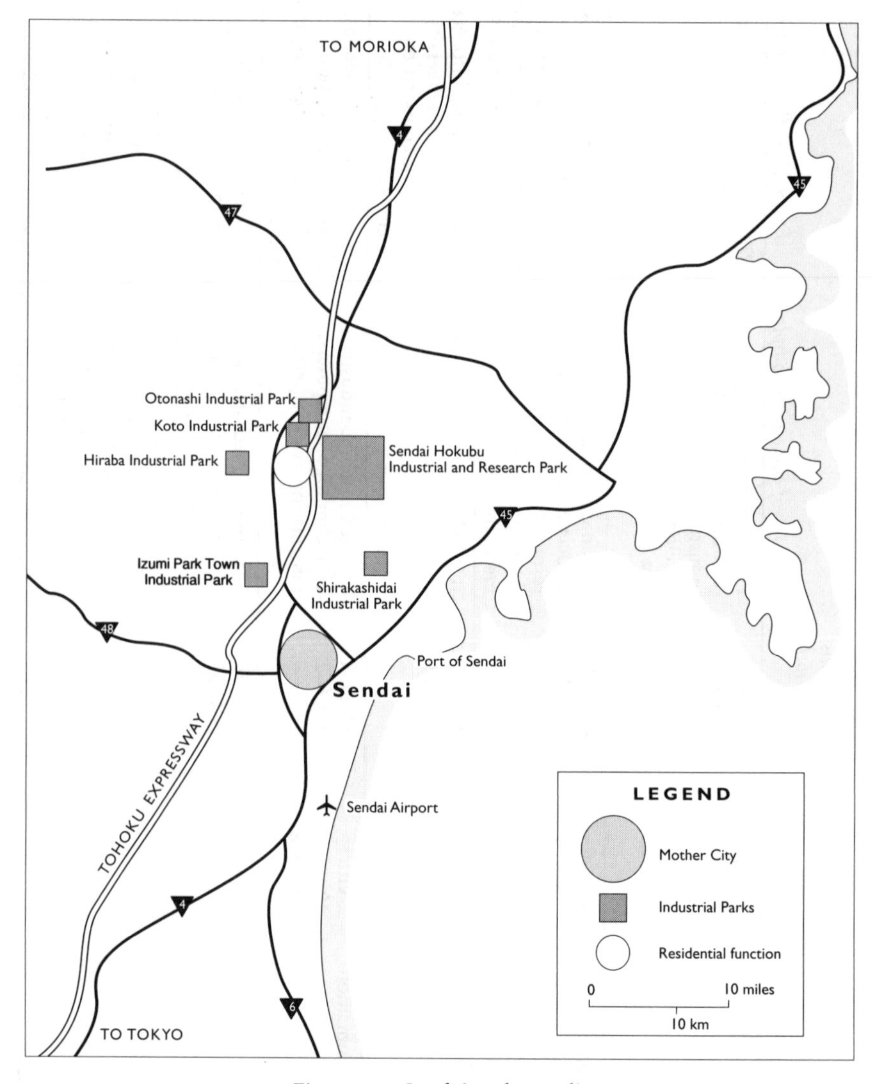

Figure 6.2 Sendai technopolis
Source: Miyagi Prefectural Government, *Sendai Research and Industrial Park, n.d.*

square kilometers) and a population of about 900,000; it is production-oriented, with a main emphasis on electronics and mechatronics, new materials, biotechnology, and urban information (Figure 6.2). Firms coming in enjoy prefectural as well as standard national incentives.

The main focus is on two sites in the attractive hilly country north of the city: the Sendai Hokubu Research and Industrial Park (Miyagi Dynamic Hills) on a 200 hectare (500-acre) site, 20 kilometers (12 miles) north of central Sendai with 40 percent sold, three companies started, and 14 others coming including Toyota and Fuji; and Izumi Park Town Industrial Park including the 21st-Century Plaza, only 10 kilometers (six miles) north of the city center. This is the research core of the technopolis: it is intended to provide an industrial support center with high technology, a center for exchanging of views and information between the private sector and academia, and a way of drawing on the vitality of the private sector. It is expected to be complete by the year 2000. The intention is that it should be developed by a "third sector" of enterprises and local governments, but the facilities up to now are mainly constructed by the private sector. It consists of a Regional Professional Training Center to provide vocational training for workers, researchers and experts of local companies, and the 21st-Century Plaza Kenko Center (incubator labs), sponsored and managed by the third sector and built during 1987–9 for tenants to rent for some 30 years, now all rented, and using very expensive equipment in the fields of mechatronics, new materials, biotechnology, and urban information. The laboratory also collects and offers information to researchers and technical experts, and offers a consultancy service. It cooperates with local university professors: eight professors from each major field act as guides and advisers. Tenants here pay below-market rents on three-year leases. The complex also has a small convention/exhibition center, and – coming in the future – specialized business services, a hotel, a trade fair, a research-based industrial park and a University of Science and Technology.

The surrounding Park Town was started in 1972, thus predating the technopolis by 14 years. It is a self-contained new town on 1,030 hectares (2,500 acres), developed by the Mitsubishi Estate Company. Designed for living and working, an important consideration in attracting private companies, it will eventually house 50,000 people in 13,500 homes (of which 10,000 will be single-family), the largest privately-built community in Japan. Currently most of its residents still commute to Sendai. But its 150-hectare (370-acre) industrial park, started in the early 1980s, is four-fifths complete, with 36 companies already occupying factories and warehouses, and others on the way; they include Motorola, Toshiba, Toyota, and other companies, especially in electronics and new materials. A golf course was completed five years ago; there are also tennis courts. Fifteen percent of the land is devoted to recreation, and by Japanese standards the environmental quality is very high.

The reasons for siting the new town were the nearby expressway and Route 4; the southerly orientation; and ease of land purchase, in the open market and without eminent domain. The original plan was for an industrial park and central office complex; but, when the technopolis was designated, the plan was changed to take advantage of the subsidies. The research center, including the 21st-Century Plaza, covers 33 hectares (80 acres) within the industrial park; the land is rented to them at a specially low rent.

Associated with the technopolis but not legally part of it, the Tohoku Intelligent Cosmos Plan was first proposed in 1987 and evolved during 1987–8, with a master plan determined in mid-1989; it is being developed cooperatively by the Prefecture, University and private enterprise to secure regional technological development. It aims to promote innovative scientific and technological research and development through the creation of a systematic institutional structure, and the foundation of some six specialized R&D companies, which are private but are largely funded by public capital. One of the first, ICR KK, founded in February 1989, is a coordinating company to facilitate industrial applications of university research and development by providing training and support services; funded 70 percent by the national Government – from the National Government Fund for Research, an independent agency established by MITI and the Ministry of Post and Telecommunications – and 30 percent by private funds, with the cooperation of 136 blue-chip companies, its president is ex-dean of the University's technology faculty. The idea is that public money is invested – effectively as free venture capital, without expectation of return – to produce patents that can be licensed; ICR can use the fees in any way it wishes, perhaps to recycle into further ventures. In August 1990 its R&D laboratories were doing fundamental research in areas like low-power communication systems, amorphous magnetic materials, rice breeding and a high-grade cold water fish laboratory; there were no commercial results as yet.

Another part of the Cosmos plan is to foster the development of "incubators": intermediate organizations designed to take basic research results and pilot them to application level in existing and nascent industry. This will include the application of research to established industries as well as the encouragement of new activities. It is "an organizational structure rather than a physical entity." As well as all this, there will be the development of a teleport, and ISDN network, a conformance and testing center, and a software module center which would establish techniques of automated software development, and the development of compatible software modules.

Nagaoka

The 130,000-hectare (320,000-acre) Shinanogawa technopolis, centered in and around the city of Nagaoka (130,000 hectares, 320,000 acres) in Niigata prefecture, looks rather more like the ideal model of a technopolis (Figure 6.3). This is the west coast, the bleak "snow country" of Japan; though the new Joetsu Shinkansen takes only 93 minutes from Tokyo, it involves the passage of the world's longest land tunnel under the Japan Alps. Here, in 1980, a new National University of Technology opened, with an exclusive emphasis on engineering and technology, and a new town plan was launched. Technopolis designation followed in 1985.

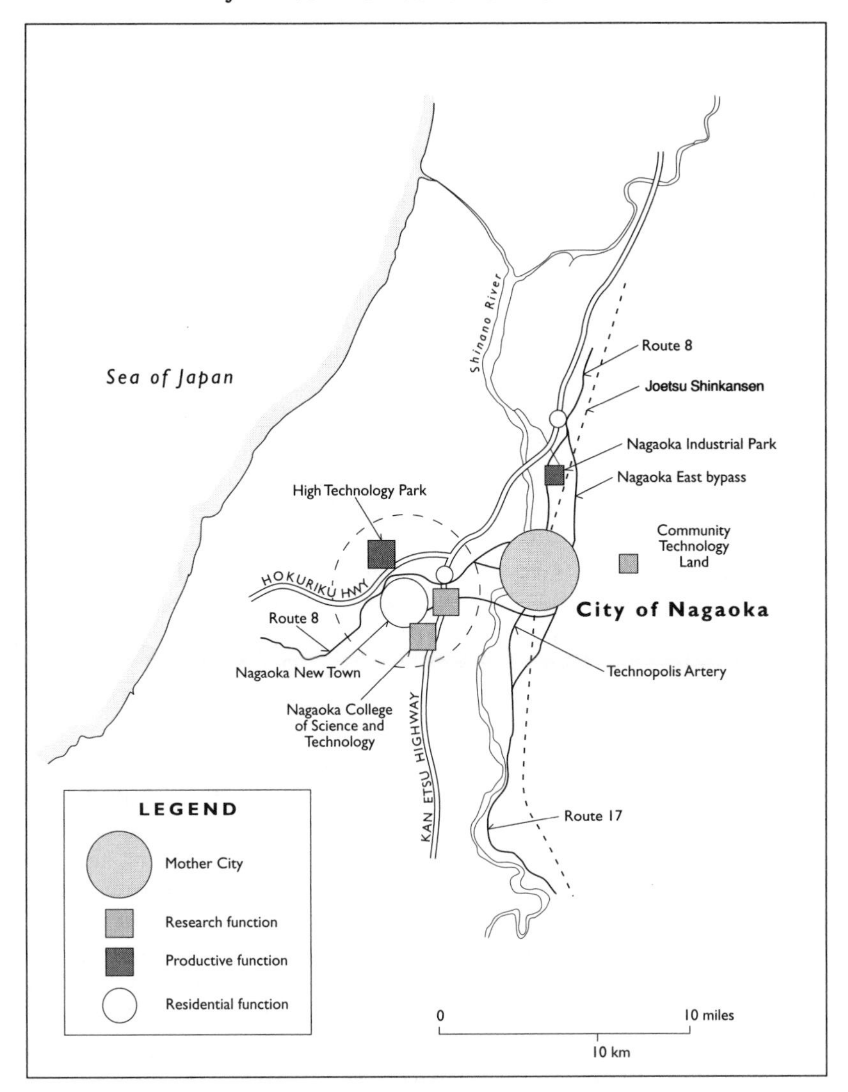

Figure 6.3 Shinanogawa technopolis
Source: Japan Industrial Location Center, 1982

It is based quite deliberately on the principle of building on an existing urban-industrial base, by injecting high technology into indigenous industrial firms. This was an old, very traditional industrial area – based on metal processing, with textiles in the neighboring towns – with problems of sluggish growth. To develop new attitudes in the old firms, the prefectural government has established an Industrial Technology Center and a Technological Coordination Center within the technopolis. The aim is to encourage outside firms in megatronics (factory automation), new materials, software and design, biotechnology, and fashion, all of which could help modernize and bring new technology into the existing industrial base. A key feature is a Research-Core Incubator Center to give companies access to a higher level of technology, to train people who will be able to train others, and to serve as a channel of information to companies. The Research-Core Center, opened in 1984, gives R&D support for products using new materials; outlets for local technologies, e.g. machining processes developed by small and medium enterprises; pollution-free car washers; and new foundry methods to rationalize processes and obtain better, harder material. It offers training through seminars on application of new technologies, with four divisions: new materials, optimized computer use, factory automation and biotech utilization. They offer on-the-job training for technicians in fashion design and telecommunications. The incubator, started in 1990, provides information and introduces consultants who explain how to start a business. They look for potentially successful ventures, and try to promote product development and marketing, starter financial help for R&D, and help with technology diffusion. The integrator core started in April 1988; all the laboratories are now used for some kind of R&D. Normally these activities are spin-offs from established companies, but they also house a company wanting to start an R&D division. The National Institute of Technology plays a critical role in providing information, in the R&D phase and in actual implementation.

They have taken care to provide the industrial infrastructure for incoming firms: 29 industrial parks were in course of development in August 1990. They could offer available labor at lower costs than in the Tokaido Corridor, cheap land and easy access by Shinkansen or expressway. They also took great care to build their technopolis around an attractive new town, which they had started before the technopolis designation. They have found it difficult to attract the right firms from outside, but have been more successful in upgrading indigenous firms through the Technological Industrial Center and the Nagaoka Institute of Technology. The late-1980s boom in the Japanese economy has benefited them: a new distribution center, built at the convergence of two national expressways, has given a boost, and at the industrial park eight companies have practically decided to come, with another six in negotiation. But the total population of the new town in 1990 was a mere 5,000, against an original target of 40,000; officials now think that they will need something more, such as a National Park or theme park, and are working with Japan's Space Agency (NASDA), with national and prefectural governments, and with local private enterprise on a space museum. This, they said, fits MITI's 1990 technopolis draft guidelines, which stress developments based on leisure and creativity.

The Nagaoka technopolis planners are realistic about the prospects for success. They stress that the payoff was very long-term: it will be 10 or 20 years before major results are evident. In other technopoles, the tendency is to judge success by the number of inward investing firms, but not here: they want a 50:50 balance between imported expertise and the encouragement of indigenous talent. And that is bound to be a long process.

Oita

Oita represents a very different model. Located in the center of Japan's southern island of Kyushu, 1,000 kilometers (600 miles) and a two-hour flight from Tokyo, isolated even from other urban areas on the island by mountain ranges, the entire Oita prefecture has only 1.25 million people. Much of it is rural: Oita city, with 400,000 people, is the only major urban center (Figure 6.4). Agriculture, forestry and fishing, which employed nearly 56 percent of the workforce in 1955, had tumbled to a mere nine percent by 1985. In terms of GDP, Oita stands 31st of the 47 prefectures of Japan; per capita income in 1985 was only 83.5 percent of the national average.

The Oita technopolis is unusual because it too is rural: it is located in a fertile agricultural area including a National Park, in a 50-kilometer (30-mile) radius around the Oita airport, which is no less than 100 kilometers (60 miles) distant from the city. Overall it covers 1,230 square kilometers (475 square miles) including four cities, thirteen towns and two villages. Oita City and Beppu City, a hot spring resort, together constitute the Mother City, though they are distant from the technopolis. Thus Oita does not meet the minimum requirement for a technopolis, which is the existence of an industrial area with highway access. Even the roads were only just about to start in 1990: they included the Airport Highway and the North-Prefecture Technopolis Road.

Oita was the brainchild of Morihiko Hiramatsu, governor of Oita prefecture in 1979, before the announcement of the technopolis concept. His idea, "one village, one-product," was that instead of subsidies, the prefecture would give technical assistance for village development. Relatedly, also in 1979, he started the airport-based Industrial Area project, to develop rural electronics and high-technology industries dependent on air cargo. Hiramatsu, an ex-MITI official, is one of the strongest promoters of foreign investment in Japan, famous for cutting red tape; wielding enormous influence, he has become known as "Mr Technopolis." His concept of a dispersed rural technopolis was at first opposed by MITI planners, but he won; MITI lacked the money for a new town.

Hiramatsu stresses four crucial attractions of Oita for foreign high-technology firms: the region's receptive environment for investment; the high quality of its labor; good access through the airport for international passengers and freight; and his personal networks, plus his knowledge of the semiconductor industry built in his days in MITI. In addition, he notes

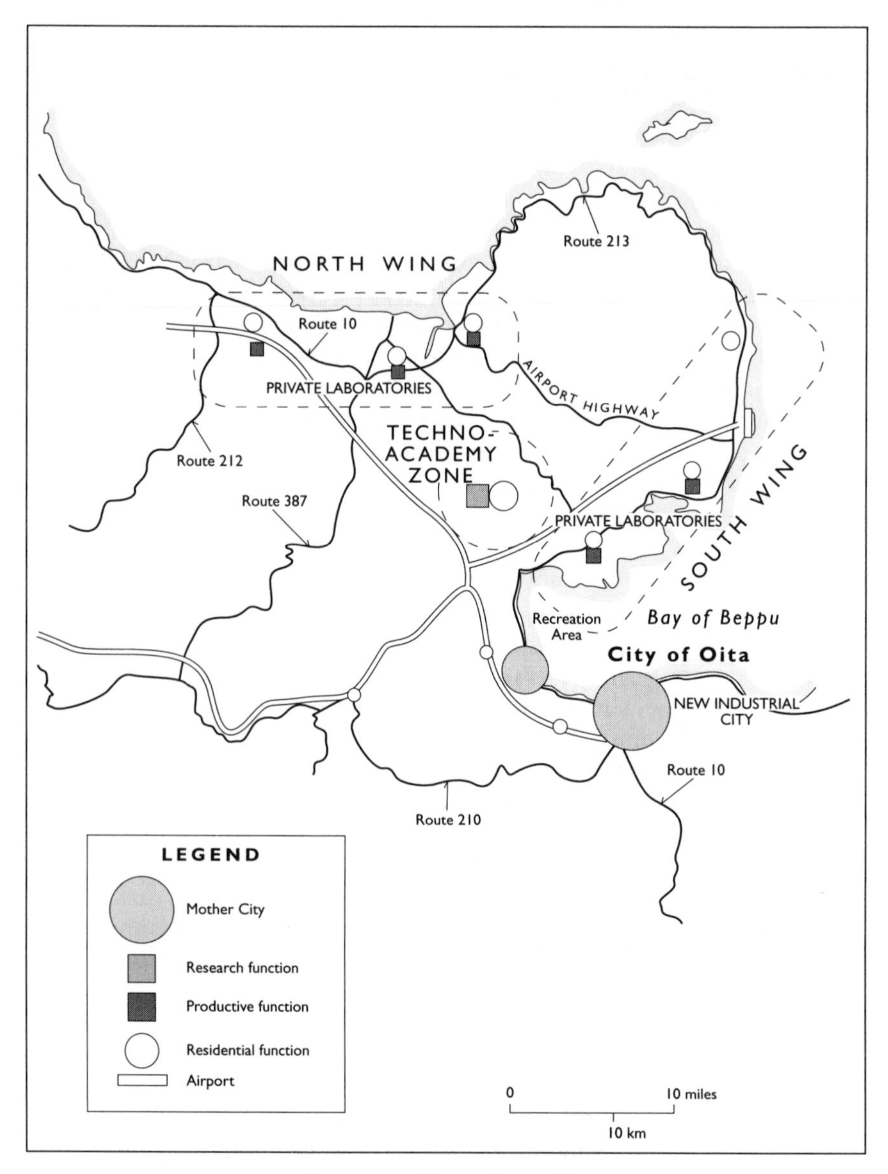

Figure 6.4 Oita technopolis
Source: Japan Industrial Location Center, 1982

that foreign companies can now receive the same benefits as any Japanese companies; for example, the Materials Research Corporation has become the first fully-owned foreign enterprise to receive a loan from the Japan Development Bank.[28] To overcome its weakness in research, the prefecture has established R&D programs at Oita University and other places.[29]

The long road from Oita City to the technopolis is a very heavily-trafficked two- or three-lane divided highway, traffic-light-controlled, lined with strip development, but with a quite idyllic quality produced by bay, mountains, and palm trees. The distance of the airport from the city is problematic, but more serious is the lack of a regional expressway or Shinkansen link. A cross-Kyushu expressway through the mountains will be ready by 1993–4, and the existing north–south highway will be upgraded, giving direct access to southern Kyushu.

The technopolis is divided into four units, each with its own small Mother City. Kitsuki, a science park next to the new airport expressway – under construction in 1990 – houses Toshiba, Oita Daihen, Ishii Tool, Hoks and TI. Oita Daihen, a branch of a parent company based in Osaka, located here because its president was born in the area; it specializes in robotics, power transformers, and arc welding equipment. The robots require software; it was difficult to get top-quality software workers in Osaka, and it was thought that well-qualified local people would want to return home. The company has 40 software engineers, all from Oita prefecture, who had worked in Tokyo or Osaka but had already returned. Recruiting between local companies is now becoming significant; because there are fewer such firms than in Osaka, more engineers are available, though not as many as formerly. Additionally, 50 percent of Oita Daihen's engineers are local graduates from Oita University who want to stay; but they might leave if living costs rise. The local city provided infrastructure and the land development; the expressway interchange was an important factor.

Oita is not the only rural technopolis in Japan, but it is the most successful. The technopolis was announced in 1982 but began operation in 1984. By 1990, 60 companies had been attracted. Within the technopolis, from 1980 to 1988, high-tech employment rose by 4,000, from 17,400 to 21,600, representing virtually the whole net employment increase in the prefecture. What percentage of this is in R&D is not clear, but 800 are employed in software alone. It is one of the most successful of all the technopoles in attracting companies, including big ones like Sony, Canon, Matsushita, Nihon MRC and Daihen Tech. (The Toshiba factory is the largest plant with 2,200 workers, and produces half the world's VLSI (1,000k) chips; it is however outside the technopolis area and was established earlier.) In all, 32 factories have been established or expanded. They make computer-related products such as large-scale integrated circuits, very large-scale integrated circuits, micro-motors, boards and other parts. They employ a total of 4,688 people; however, the two largest, Kyushu Matsushita Denhin and TI, were established in 1970 and 1973 respectively, long before the birth of the technopolis project. Subtracting them shrinks the total to 3,508, and of this only some 2,500 (0.4 percent of total prefecture employment, or perhaps 1 percent including induced service jobs) represents local people. Further, most of the work is of a routine production-line character; there is a small Fujitsu R&D facility outside the technopolis area, but local managers doubted the ability of the technopolis to grow R&D.[30]

One great advantage for incoming firms here is the cost of land, which is a mere one-thirtieth of the Tokyo level. High-quality labor from local schools was another critical attraction; it is critical for integrated circuit production, which cannot be offshored for that reason. Also crucial were the personal ties of Governor Hiramatsu, who used his prestige as a former top official of MITI to attract firms like Sony; they came as a favor, though the location was not ideal.[31]

Yet there are remaining problems. It is difficult to find evidence of a coherent strategy: there is merely a collection of factories strung out along Oita Bay, most built before the technopolis project began in 1984, as part of the earlier airport-based industrial development scheme. Transportation links with Oita City were still poor in 1990, though air links to Tokyo and Osaka were good. But air links are not nearly as important as was thought by the original technopolis committee. A survey shows that more than 80 percent of integrated circuits are transported by truck, and that the proportion is rising. The Toshiba plant, only a half-hour drive from the airport, sends half its output by truck. The Canon plant, which used to send parts and products by air, now sends less than 10 percent. And, here as elsewhere, proximity to highway interchanges has increased in importance as a locational factor – as was very evident in the summer of 1990.

More importantly, in Oita one crucial ingredient – support and involvement of universities – appears lacking;[32] cooperation between companies with Oita National University has not been obtained, and combining the R&D element with production is not easy. There have been few cases of technology transfer; local industries have not moved beyond making parts for IC chips, and it does not prove easy to decentralize R&D.[33] Oita therefore scores high on MITI's production (*san*) axis but low on educational (*gaku*) and residential (*ju*) aspects, since it has done little so far to draw people back from big cities.[34]

Oita can be compared with Kumamoto on the opposite coast of Kyushu, since both are successful cases of technopolis development, equally renowned for their production of semiconductors. In fact, the central half of the island, which includes both zones, began to attract semiconductor manufacture from the late 1960s because of low land and wage costs: NEC and Mitsubishi Electric came to Kumamoto, TI to Oita, and so on. Kyushu now has 25 IC makers with 10 percent of world production (25 percent of Japan's) and is known in Japan as "Silicon Island."[35] But in contrast to Silicon Valley, where chip makers were typically start-ups financed by venture capital and had close links to local universities, the two regions in Kyushu are without exception production facilities owned by big companies based elsewhere, whose function is merely to execute designs drawn up in Tokyo or Osaka. Thus, the arrival of those chip makers did little to alleviate

Kyushu's brain drain.[36] If R&D does move, it goes to "Silicon Road," the northern half of Honshu including Iwate, Miyagi, Fukushima and Yamagata prefectures, which offer the best trade-off between access to Tokyo and low land prices.[37]

Kumamoto

Kumamoto technopolis, on the west coast of central Kyushu across the mountains from Oita, is a 956-square kilometer (370-square mile) "Techno-Corridor" containing 739,000 people, embracing the local airport, two cities, twelve towns and two villages. The main urban functions, like retailing and culture, are concentrated in Kumamoto City, an old castle town which is now a thriving regional metropolis and the technopolis Mother City; research institutes and production functions are next to the eastern and northern borders of the 10–20 kilometer (6–12 mile) wide Techno-Corridor, which follows the belt expressway and Highways 57 and 387 east and north-east towards the mountains (Figure 6.5). Most of the area is still agricultural. Within it, several high-tech "forests" will be scattered along the highway in a high-quality living environment; America's Route 128 and Research Triangle provided the model.[38] Implementation began in 1984; the infrastructure is only now being built, so that in the early 1990s the project was only just gathering momentum.

The background, and the baseline against which the technopolis project has to be judged, has already been noticed: even before 1980, Kumamoto had become a major center of Japanese integrated circuit production. No less than 40 percent of all Japanese IC output comes from Kyushu, of which half (i.e. 20 percent) is from Kumamoto prefecture. NEC and Mitsubishi are both located within the technopolis zone, with integrated microchip factories and subcontractors. Several factors originally attracted the industry here: pure fresh underground water, clean air, hard-working people, a forward-looking prefectural government, and excellent centers of technological learning and research.[39] As an industrial late-starter, Kumamoto has certain advantages: it has no old, obsolescent equipment or social systems, and it has greater energy. Governor – later Prime Minister – Morihiro Hosokawa is a person of great energy, fully in touch with technical developments worldwide.[40] Manufacturing employment rose by 39 percent between 1970 and 1984, and the value of shipments rose five times. But, as the industry has matured, it has shifted toward more skilled male employment: NEC, perhaps the largest IC plant in the world, moved from 90 percent female employees to 50 percent between 1970 and 1985. The women were left as semiskilled production operators, and the technopolis program does not provide for upgrading their skills; and, as industry offshores its operations overseas, female workers are likely to be the first to feel the effect.[41]

The region contains several universities and colleges with excellent

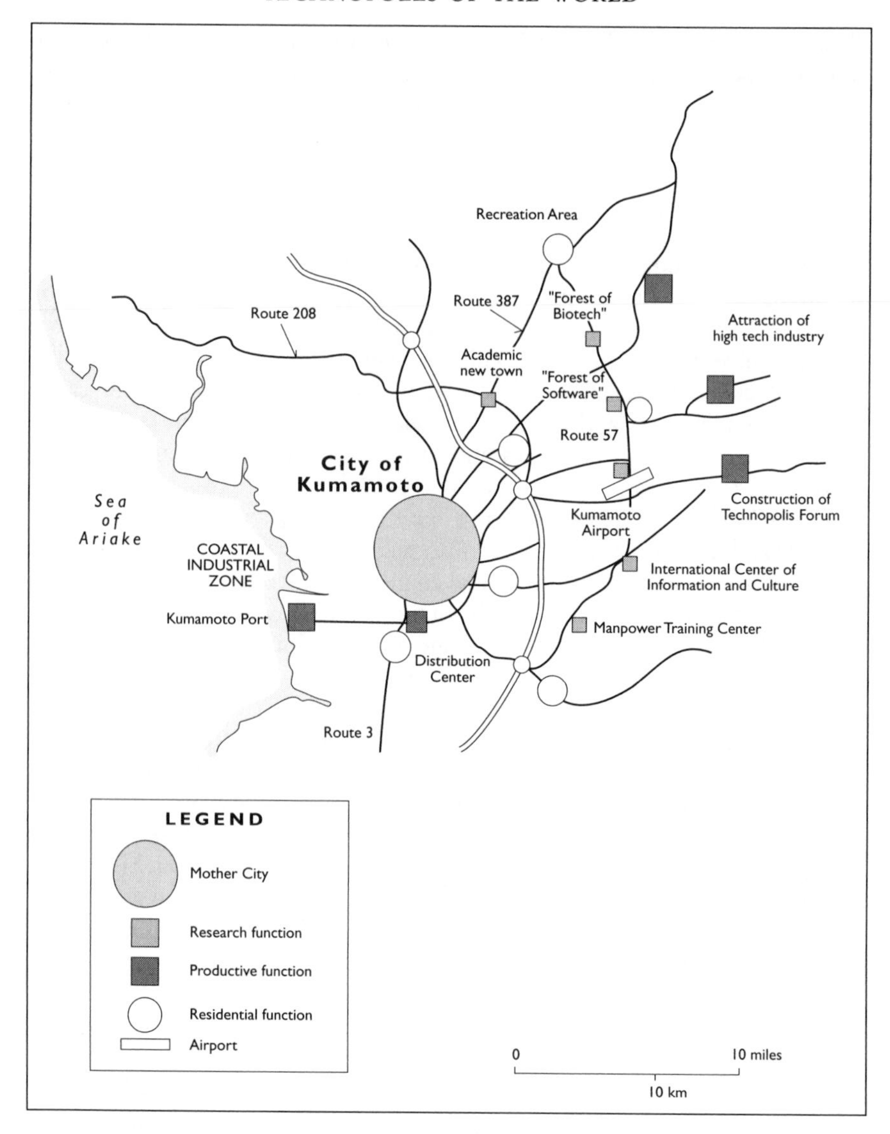

Figure 6.5 Kumamoto technopolis
Source: Japan Industrial Location Center, 1982

engineering graduates, most of whom move to employment elsewhere, but who are increasingly trying to return because of overcrowding and high housing costs in Tokyo and Osaka; the Kyushu prefectures have opened offices in Tokyo and Osaka to help bring them back.[42] The technopolis plan recognizes these shifts; it emphasizes four priorities – automation, biotechnology, electronics, and information systems (software) – and tries to implement them through local incentives including low-interest loans, subsidy on real estate acquisition and corporate tax, prefectural tax holidays, special new enterprise potential loans, and infrastructure subsidies to local authorities.[43] The industrial policy aims both to introduce bigger outside firms and to promote small local firms; the emphasis now is on how the big companies can somehow help develop the local firms not merely through subcontracting, but also through exchange of knowledge.

The plan is to develop not just production but also research functions, hence the construction of a research park in the center of the planned corridor together with provision for housing, culture, arts, and information. And, in contrast to Oita, Kumamoto has evidently forged a strong link between local government, industry, and academia. The best evidence is found at Kumamoto's Techno Research Park, the core of the technopolis development.[44] Completed in April 1991, it is the symbol of the technopolis and its functional center; it will mainly be involved in software production. It stands on Takayubaru Plateau, a green plain at the western foot of the mountains surrounding Mount Aso, close to the airport and 20 minutes from the city by car via the Second Airport Access Highway. Amid large areas of green space, the complex includes the Kumamoto Technopolis Center, the Applied Electronics Research Center, and the Kumamoto University Cooperative Research Center. Further lots will be sold. The prefecture proposes a second research park in another location, not yet determined, to promote production functions.

Visitors to the Center arrive in a high-tech lobby to see a promotional video that stresses the congestion and strain of everyday life in Tokyo; Kumamoto is only 90 minutes away by air, with rural tranquillity, natural beauty, culture and great development potential – cheap land, and skilled labor. They are shown remote central databases that can be accessed over phone lines from the center or from outside (but only at long-distance phone rates); a local area network was just becoming available within the research park but not outside it. Videoconferencing with other centers is also possible, but only on advance order.

The Center is trying to promote exchange with the three local technical universities and two technological high schools, as well as with 60 enterprises, through a Promotion Club for Science and Technology; meetings are held in various locations, in a relaxed social atmosphere. This is a deliberate experiment in the informal exchange of information, and is not yet completely developed; speakers come from the universities, companies and administration, and a journal is published. Another club was recently established just for companies based on the park.

The first priority, clearly, is to attract outside companies. The critical factors are the quality of local labor, the good environmental quality, and the living conditions for returning locally-born people. Lack of housing will be remedied by development around the Research Park. There are major financial incentives: as in all technopoles, national taxation on land and equipment is relaxed, the precise rate of relief being left open for case-by-case negotiations with the national tax authorities; the depreciation rate is halved; and land is readily available at less than market value, for a negotiated price that varies greatly. Here, a significant factor is that the area is all deregulated farmland which cannot be sold in the regular market, only through a public body.

Later, the stress will be on improving the technical level of existing indigenous firms; this will take more time. Three-and-a-half thousand computers have been placed in schools within the prefecture. The Center will also encourage new start-up companies. The Applied Electronics Research Center will guide technological development and carry out co-operative development with local firms. A foundation will guarantee loans from banks. And the Technopolis Center will train young people, just graduated from high school, for two years in management techniques, after which they will all get jobs in local firms.

The Advanced Electronics Research Center, housed next door in the same campus-like complex, is jointly financed on a 50:50 basis by the public and private sectors. It offers small companies access to specialized resources, such as powerful computers and expensive testing equipment. In addition, a permanent staff of 10 researchers, all returnees to Kumamoto from jobs with large companies, are on hand to work with local engineers to develop applications of technology. As a result of these efforts, it is reported that several hundred Kumamoto-born engineers have come back to live in the prefecture.[45] Local firms rent space here, but simply in order to enjoy cheaper rents than in the city; synergy is not even in question, we were told.

Nearby is Kumamoto University's Cooperative Research Center. Started in 1983 under a Ministry of Education program, and located in a building completed in 1988, it aims to pursue cooperative research with industry, help industry through education programs, provide technical advice to companies, and pursue cooperative research with other local universities. The main fields are biotechnology, environmental science, new materials, electronics, and medical equipment. Most projects, when we visited, involved large Kyushu-based companies. Output is patented through a public agency, with joint ownership between the university and the company, but the company enjoys a seven-year sole right to exploit the patent; the center is very rarely involved in generating start-up companies.

The technopolis plan called for industrial employment to increase by 30 percent between 1980 and 1990, output by no less than 130 percent.[46] Officials felt in 1990 that employment generation was only just beginning to happen – though they argued that local people had a new confidence in the future. But details were available for some 14 firms, almost all Japanese, established in the technopolis since 1984, which had generated just over 392,000 jobs. They included four local firms with a total of 703 jobs, giving an average of only 176, plus one for which no details were given; nine Japanese national firms with a total of over 391,000, giving an average of 43,400; and one foreign firm producing 184 jobs. Many of the moves had been very recent. A substantial proportion of the new firms are in software. Clearly, officials said, for the technopolis policy to succeed, introduction of outside software houses is not enough; it is more important to develop a manufacturing base supported by software.

EVALUATING THE TECHNOPOLIS PROGRAM

There are two ways of evaluating the technopolis program. First, in terms of crude numbers: how many new factories have been created, and how many people have they employed? And, to probe a little deeper, how many of them are simply branch-plant operations controlled by companies in other cities? Second, and related to this last, is the deeper structural question: how far has the program succeeded in creating genuinely new innovative capacities within the technopolis regions?

In first looking at the numbers, there is one major obstacle: surprisingly, though MITI claims that industrial investment rates are higher in the 26 areas than in Japan as a whole, it releases no regular figures of the results of the program. There is however some partial information (MITI, 1988a, 1988b, 1990; Nakano, 1988; Yamasaki, 1990). In the first 14 technopoles, authorized in 1984, the annual average number of new establishments during 1984–7 was 1.5 times as large as that of 1981–3, against a national increase of 1.3 times.[47] Similarly, for 1984–8 as against 1981–3, factory openings were 1.5 times greater (MITI 1990: 11). Even so, during 1984–7, the factories located in technopoles totalled only 6.4 percent of all new factories in Japan.[48]

Detailed work by Sakamoto shows that though computer-based facilities are still concentrated in the three largest metropolitan regions (Tokyo, Osaka, and Nagoya metropolitan areas), there has been a decided decentralization trend during the 1980s. The original 14 technopolis prefectures had 10 percent of information-processing workers in 1981, increasing to 12.9 percent in 1985; the Tokyo metropolitan area's share dropped from 61.1 percent to 55.8 percent, that of Osaka metropolitan area increased from 16 percent to 18 percent. Thus there is a shift of employment from Tokyo to other areas.[49]

This process of decentralization is confirmed by changes in the number of establishments. Tokyo's share of establishments drastically dropped from 58 percent in 1981 to 32 percent in 1985 in the software industry, and from 30 per cent to 23 percent in the information-processing industry. Osaka increased its share from 16 percent to 24 percent in software and from 11 percent to 14 percent in information processing. In the 14 technopolis prefectures the growth was dramatic: establishments increased by 425 percent in software, and by 34 percent in information processing. Since the percentage of software branch offices decreased, while employment grew more slowly than firm numbers, the implication is that growth in the technopolis prefectures came from small and medium-sized independent firms.[50]

Further, information-related businesses in the technopolis areas are increasingly exporting their products, particularly to Tokyo. The 14 prefectures produced 7.9 percent, and consumed 8 percent of national output in 1981; they produced 9.4 percent, and consumed 9 percent in 1985. Thus supply began to exceed demand during that period; the percentage of sales made within the prefecture dropped, while the share sold to Tokyo rose.[51]

The most recent and comprehensive quantitative analysis is by Yamasaki (1990). His conclusions are decidedly negative. He uses three indices: value of manufactured goods output; manufacturing employees; and population. Superficially, the 26 areas slightly increased their share of all three between 1980 and 1985, and on output from 1985 to 1987 (Table 6.2).

But the picture looks very different as soon as we distinguish between two groups of technopoles. The first includes three areas within 300 kilometers (200 miles) of Tokyo – Utsunomiya, Hamamatsu, and Toyama – which would have grown anyway as Tokyo's economic sphere expanded, and another five areas that were growing already: Koriyama, Kofu, Asama, Yamagata, and Sendai-Hokubu. Excluding these eight, the remaining 18 areas did not perform well at all (Table 6.3).

All of the three indices declined from 1980 to 1987. And a substantial proportion of these 18 areas were performing badly on one or more of the three indices, as Table 6.4 shows. Fully half of these technopoles recorded an absolute decline in manufacturing employees from 1980 to 1987; two,

Table 6.2 26 technopolis areas: three indices

	Share of national total		
	1980	*1985*	*1987*
Output value	9.41	9.54	9.55
Employment	10.26	10.34	10.34
Population	11.11	11.17	11.16

Source: Yamasaki, 1990

Table 6.3 18 technopolis areas: three indices

	Share of national total		
	1980	*1985*	*1987*
Output value	5.70	5.38	5.20
Employment	5.44	5.31	5.21
Population	7.07	7.08	7.05

Source: Yamasaki, 1990

Table 6.4 18 technopolis areas: number with declining indices

	1980–5	*1985–7*	*1980–7*
Output value	1	5	1
Employment	6	10	9
Population	0	4	2

Source: Yamasaki, 1990

including Oita, experienced a decline in population. Further, Yamasaki argues, the effects within each region may have been equally negative: since many designated "Mother Cities" were prefectural capitals, the programme may actually have aided the concentration of economic activity there, rather than spreading it into the surrounding region.

Figures do not tell the whole story; a deeper question is whether the technopoles are essentially still branch-plant economies dependent on Tokyo or Osaka, or are developing some indigenous innovative growth potential. Here, independent analysis – both by Japanese and American observers[52] – is far from sanguine:

1 *Failure to achieve original vision*. Edgington argues that the program failed to achieve its original objective of a satellite city integrating R&D, educational facilities, and production facilities for high-technology industry. Some are conventional satellite new towns, including those already started by the Japan Regional Development Corporation in the 1970s. Many technopolis areas depend on the pull of cheap land and skilled labor compared with the big cities.[53]
2 *The "branch-plant" syndrome*. Further, decentralized firms have been mainly confined either to the making of parts for shipment to Tokyo, Osaka or overseas, or to routine assembly; so very little technology transfer has occurred between incoming factories and local industries. It is for this reason that most prefectures have been forced to build, or plan to build, their own "Technopolis Center" to introduce state-of-the-art technology into the region.[54] Nishioka argues that the decentralization of high-technology industry associated with R&D will consist mainly in simple parts assembly functions, and R&D oriented toward improvement of existing processes. Because of this, a major effect upon the regional economy, including technology transfer, is not to be expected.[55]

   Sakamoto reaches a similar conclusion. He thinks that during the 1980s the Tokyo region simply could not keep up with the demand for information-technology products, because of labor shortages. It was this that impelled the birth of establishments in technopolis areas, including both branch plants and independent firms. But the latter were not necessarily started by local entrepreneurs; many are subsidiary firms of large-scale corporations headquartered in Tokyo. They create jobs, but they may well compete with indigenous firms for skilled workers, weakening their competitive position. And, since they tend to subcontract for their parent companies, it is doubtful how far they help raise local technological levels.[56]

3 *Failure to develop university–industry links.* Close links between local universities and industry, as in Cambridge and Stanford Universities, are virtually unknown in Japan because of restrictive regulations and the inability of academic staff to evaluate any work of their peers conducted outside the university campus. Thus, although local universities are involved in technopolis activities, it is the prefectural technical laboratories that provide the vital link between university research and industrial application.[57] In any event, the program makes no attempt to address the agglomeration of premier universities and corporate headquarters in the Tokyo region; there will be no relocation of these key functions, and thus of the key creative professionals in them.[58]
4 *Lack of "soft" infrastructure.* Prefectural governments seem to have concentrated on construction of "hard" infrastructure – roads, airports, university and laboratory facilities, technology centers, and industrial/research parks – and too little on the "soft" infrastructure of R&D consortia, venture capital funds, and university research needed to drive the technopoles. In most cases there was no "magnet" infrastructure or "leading edge" research technology which could by itself attract or retain footloose high-technology firms. A sole exception is the Nishi-Harima technopolis in which the construction of a synchrotron radiation ring is planned.[59]
5 *The failure to move R&D.* Major business corporations have been reluctant to locate their research facilities in provincial areas outside big cities.[60] Yamasaki shows that for industrial laboratories established since 1981, proximity to headquarters has replaced proximity to the factory as the key locational factor. And, since 1984, a new spatial division of labor has emerged: between factories producing prototype models, called "mother factories," and factories producing standardized models. Thus NEC conducts basic research and circuit design in Kawasaki City close to its Tokyo headquarters, and develops mass production techniques close by at Sagamihara City, while undertaking the actual mass production through its subsidiary, Kyushu NEC, at Kumamoto.
6 *Lack of inter-industry linkages.* A key issue is the linkage of high-tech industries to the local economy. It appears that pure R&D facilities do not need to form local linkages, while once the technology is commercialized as large-volume, standardized products, they may not generate local linkages. Only firms producing custom products are likely to forge local input linkages and create downstream local markets.[61]
7 *Lack of spin-off.* Similarly, firms producing pure research, one-of-a-kind products, or standardized products may have limited potential to create or encourage spin-offs. Here, the parent firm gains few benefits from the

creation of spin-offs, and may incur considerable costs when personnel leave their employment. In contrast, custom producers, which buy parts in different batches from a variety of vendors, may encourage spin-offs, particularly if the spin-off's product is tailored to the parent's product line.[62] It seems indeed that the plan was based on somewhat naive "trickle-down" theories that may fail to materialize, because spin-offs will not occur from outside firms to new local ones.[63] For instance, one key business which it is hoped will develop as a "cottage industry" is that of software writing, but this ignores the fact that most small Japanese software houses that do well are either bought out by large electronics firms or are overwhelmingly dependent on a single large client.[64]

8 *Failure to attract key workers*. At present, 60 percent of the information technology workers are employed in small and medium-sized firms in the Tokyo region. Given this, the technopolis program may fail to attract skilled labor, thus causing a problem of labor shortage.[65]

9 *Fiscal burden on local governments*. There is a danger that competition among the prefectures will result in a game of regional Darwinism, in which some will survive while others will be crushed.[66] Prefectures may find it difficult to persuade the private sector to invest in activities stimulating domestic demand, such as construction, housing, real estate, and transportation, because these are no high-tech sectors with the promise of high growth and high profits. So the public sector will probably be asked to shoulder most of the infrastructure costs.[67]

10 *The continuing challenge of offshoring*. In addition, the high value of the yen against the US dollar since September 1985 has undermined the technopolis strategy, by forcing firms to hollow-out and offshore their production operations to less expensive sites in Asia, Europe, and the United States, thus reducing the flow of technology from Tokyo and Osaka to technopolis sites.[68]

11 *An overall conclusion*. It is difficult to avoid the two main conclusions that emerge from these studies. First, the technopolis program has not been very successful in generating new activity in the majority of cases. Yamasaki concludes that there will emerge an increasing division between the eight technopolis areas within 300 kilometers (200 miles) of Tokyo – Utsunomiya, Kofu, Hamamatsu, Toyama, Koriyama, Asama, Yamagata, and Sendai-Hokubu – and the rest. Prospects for the first two, which are within 100 kilometers (60 miles) of the capital, are particularly bright because they should attract laboratories and "mother factories." Of the other 18, the Nishi-Harima technopolis in Hyogo prefecture will do well because Harima Science City, its core, is the site of construction of one of the largest synchrotron radiation storage rings – a high-energy physics testing facility – in the world. On completion in 1998, it will attract research laboratories in new materials, ultra-large-scale integrated circuits (LSI), and biotechnology. For the remaining 17 sites, Yamasaki

> concludes that prospects are poor. A "Technopolis 2000" committee, set up in 1989, is said to be working on a strategy of encouraging tourist developments to take advantage of proximity to airports.

The second conclusion is that even the limited successes mainly lie in promoting branch-plant type operations, which have little innovative potential and are highly vulnerable to the risk of offshoring. And this points to the underlying tension, or contradiction, between the two main objectives of the program: national development and regional development. The original technopolis programme stressed the need to locate in areas offering environments that were favorable to R&D; and Professor Ishii, one of its architects, admitted even then that as a policy of national technology development, it need not necessarily be implemented in provincial regions. The irony here is that the technopolis program has actually stimulated local governments in the capital region to fight back with plans to develop R&D functions. As will be seen in Chapter 7, Kanagawa prefecture has launched its "Brain Center Programme," including "Kawasaki Maicon City," "Kawasaki Technopia," "Kanagawa Science Park," "Atsugi-Morinosato," "Kohuku New Town," "Hakusan Hightech Park," and "Ken-ou Hightech Park," with at least another nine plans in course of formulation.[69] So, Yamasaki concludes, while failing as a regional development program, the technopolis program has successfully encouraged further R&D agglomeration in the Kanto (Capital) Region. In other words, ultimate irony, it has successfully created a "Tokyo–Yokohama Mega-Technopolis."

What is perhaps most surprising is that MITI appears to have positively encouraged this development. Since it launched the technopolis program in 1980, it has followed up with projects carrying exotic titles like "Research Core" and "New Media Community," while other ministries have countered with programs like "Teletopia," "Intelligent City," and "Greentopia."[70] Some of these have a decided spatial bias towards Tokyo and the Kansai (Osaka–Kyoto) area, and so appear directly to contradict the objectives of the technopolis program. But perhaps this is not so surprising: MITI's general approach to industrial policy, during the 1980s, appears to combine very steadfast long-term strategic objectives with a remarkable capacity to scatter-shoot all kinds of varied, sometimes even contradictory, implementation tactics. In 1980 no one in the world knew how to promote innovation in lagging peripheral regions. By a subtle combination of local initiative and central assistance and guidance, MITI was going to try to find out. But perhaps it would also hedge its bets, because nothing could be allowed to obstruct the central aim of achieving and maintaining global technological leadership.

This however underlines a more general point: the technopolis policy cannot be seen in isolation. It represents a very deliberate attempt – almost certainly, the most determined yet made by any major industrial nation – to pursue a concerted innovation-based regional policy. But it makes sense only in terms of an even higher national priority to shift the Japanese industrial base, away from exploitation of imported technologies, and toward a world role in the development of leading-edge high-technology industry. What is still uncertain is how far these two policies can be made compatible. Other things being equal, an emphasis on innovation may well fortify the tendencies toward centralization that are so evident in the Japanese urban and regional system. The regional policy emphasis may try to harness the forces working in the opposite direction – particularly the massive negative economic externalities that operate in the Tokyo area – but the result may merely be the development of a branch-plant R&D system, in which regional laboratories take their orders from Tokyo headquarters.

MITI would doubtless counter that by saying: time alone will tell. Certainly, the early 1990s are too early to reach a definitive judgment on a program that was first announced in 1980, passed into law in 1983, began to operate in 1984–5 and is planned to achieve full operation in the first decade of the twenty-first century. What can be said is that to be truly successful, the program will have to achieve a revolution in Japanese thoughtways and customs: the development of an American-style system of strong relations between local universities and local grassroots entrepreneurs, with spin-offs from university research into new companies, which the entire structure of Japanese professional and social relationships would seem systematically to discourage. What then remains unclear, in the technopolis experiment, is how a bottom-up regional policy can be grafted on to one of the most top-down centralized systems of industrial management in the world. But the Japanese have surprised the rest of the world many times in the late twentieth century, and may well do so again in the early twenty-first.

7

THE METROPOLIS AS INNOVATIVE MILIEU

The argument now comes to a disturbing provisional conclusion: despite every heroic attempt to the contrary, from Silicon Valley onward, the sources of technological innovation proved remarkably resilient to all attempts to shift them. Silicon Valley was in fact the great historical exception, simply because it was new; for the first time, an innovative milieu was created by deliberate human action, in a place that seemed to have none of the normal prerequisites. In sharp contrast, for some two centuries previously, the historical record shows that technological innovation had come out of two rather different kinds of city. The first was a rising city on the borderland of the then economic cores of the world, enjoying certain advantages of propinquity to the existing centers, but at the same time free of their inherited traditions and constraints. Manchester in 1770, Detroit in 1900, Glasgow in 1950 were examples. All these, it needs stressing, were quintessential laissez-faire cities: their rise had logical origins and was associated with particular individual actors exploiting particular structural conditions at particular points of time, but it owed nothing to any deliberate planning process.

The second was the existing major metropolitan city, where skilled craft traditions allied with new demands coming from state or commercial power, in turn creating demands for scientific advance and for military prowess. Berlin between 1880 and 1914 was a classic example of this, and was indeed the Silicon Valley of its day. But some of the same characteristics were shared by London, Paris, and New York in the nineteenth century, and by Tokyo in the twentieth.

Of this group of five world cities, Berlin lost its position after 1945, purely as an outcome of world politics; Munich, until then a relatively under-industrialized city, was the chief beneficiary. And the Greater New York region failed to maintain the head-start that sustained it until 1945, losing its accumulated advantages to other, sometimes upstart, metropolitan centers – among which, the Southern California region, based on Los Angeles, was the most aggressive. But the remaining great metropolitan innovative milieux – London, Paris, Tokyo – are worth study, for two reasons. First, because in varying degrees they did continue to develop their high-technology industries, adapting to changed circumstances by developing new industrial traditions out of older ones: in the 1880s and 1890s precision engineering begat electrical engineering; in the 1960s and 1970s electrical engineering begat electronics. Thus, they have continued maintain their positions as the dominant high-technological industrial centers in their respective national economies. And second, because as they did so, they in effect changed their internal economic geographies: beginning as tightly-bounded artisanal quarters in the inner city, they spread into highly-decentralized industrial corridors or belts – London's Western Crescent, the south-west quadrant of Paris, Kanagawa prefecture south of Tokyo. In both these respects, they offer unique lessons in the dynamics of innovative growth.

However, as everywhere else in this inquiry, there is a deeper and finally central question: in what sense do these giant urban complexes represent genuine innovative milieux? What synergies do they generate between the establishments and institutions that are locked together in them, such that the tradition of innovation constantly renews itself from decade to decade and from generation to generation? Without doubt, in origin they all had this characteristic; the question is how far, in the shift from small-scale artisanal production to large multinational corporations, they have managed to retain it. If we can answer this question in these particular geographical circumstances, it will help us understand how to create such a milieu in other kinds of locale.

In this chapter, therefore, we will first look at this group of three metropolitan cities. Then we will turn to the two upstart cities that emerged in the mid-twentieth century, to ask how they so suddenly and so successfully eclipsed their older-established rivals.

THE METROPOLITAN SURVIVORS

LONDON

The western side of the London region – variously described as the "Western Corridor," "M4 Corridor," "Western Crescent," and "Western Sector," and comprising a few counties immediately southwest, west and northwest of London, between about 20 and 50 miles (35–80 kilometers) from Piccadilly Circus – is outstandingly the high-technology core of the British economy (Figure 7.1). That statement must be set in context: Britain, like Europe generally, is no longer one of the great technological crucibles of the world. The Western Sector's performance is exceedingly modest when compared with that of some of the great innovative milieux, as Tables 7.1 and 7.2 demonstrate: the area is exceedingly small in employment terms, and its growth was extremely slim. Astonishingly, too, British

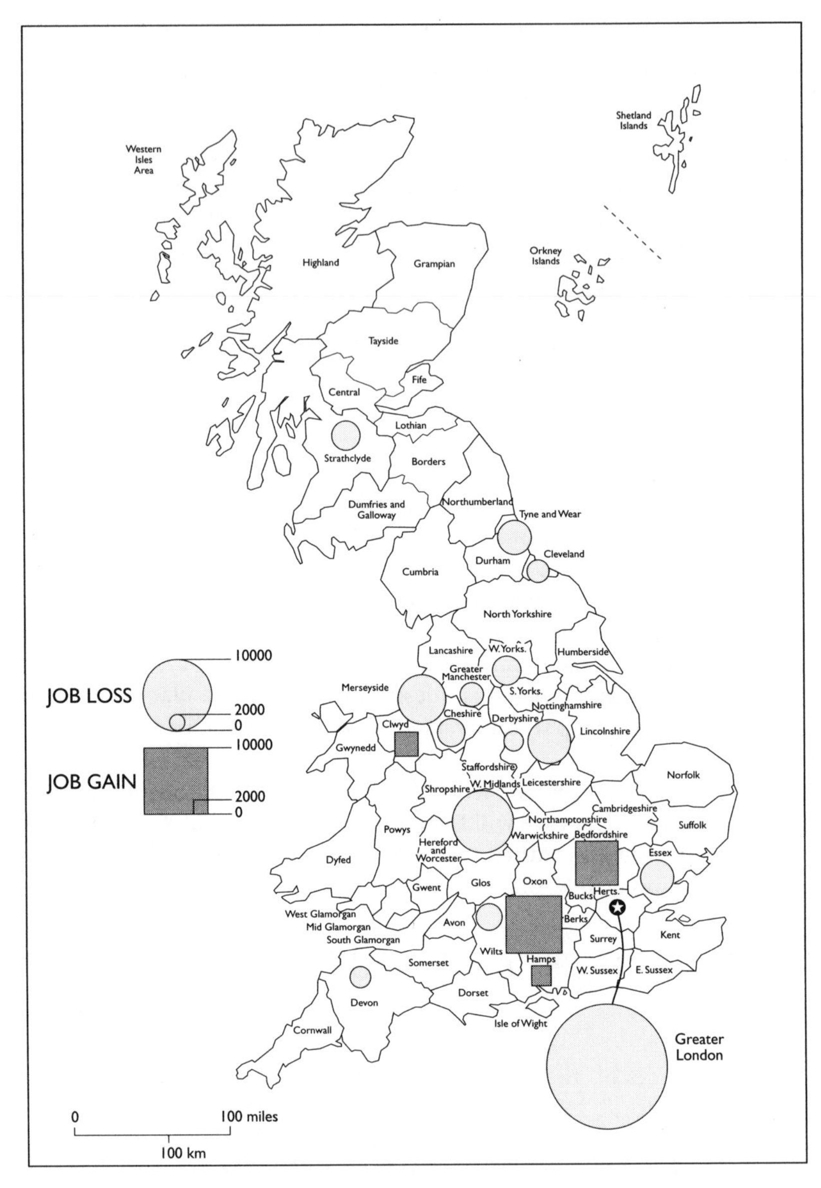

Figure 7.1 Great Britain: high-technology industrial change
Source: Annual Census of Employment, Great Britain

high-technology employment was actually declining overall in the late 1970s; even within the South-East region, the gain in the Western Sector barely compensated for the huge losses of employment within Greater London. None the less, London's Western Sector does represent a remarkable case of long-continued innovative momentum; and for that reason, it is important to study.

Its origins, and its continued growth, owe little to deliberate, planned actions in the conventional sense, but do exhibit conscious volition in a wider context. The beginnings lay in small artisanal workshops in certain artisanal quarters close to central London, at the end of the eighteenth century, which earned a living by the craft production of highly specific scientific instruments and precision machinery. One of them, around Clerkenwell immediately north of the City of London, specialized in watches and clocks until it was almost destroyed by Swiss competition during the course of the century. Another, more relevant for this story, was located in Pimlico, close to the seat of government in Westminster, spilling out from there into Lambeth immediately across the river.

Though this complex has never been adequately analyzed, it seems clear that it had close relationships to the new scientific-technological institutions that were then being established with headquarters close to Parliament, such as the Institution of Civil Engineers and the Institution of Mechanical Engineers. It was located on marshy ground, which offered cheap sites; the South Bank of the Thames, from Vauxhall Bridge to London Bridge, was then the greatest concentration of engineering shops, large and small, in all

Table 7.1 Great Britain: concentrations of high-technology industry, by county groupings, 1981

	Employment	*Location quotient*
London Western Crescent		
Greater London	91,400	0.85
Hertfordshire	45,100	3.60
Hampshire	33,800	2.00
Berkshire	19,700	2.04
Buckinghamshire	6,900	1.15
Surrey	18,200	1.79
Total	215,100	
North-West England		
Greater Manchester	27,300	0.86
Lancashire	30,000	2.01
Total	57,300	
"Silicon Glen" (Central Scotland)		
Strathclyde	23,600	0.88
Lothians	9,000	0.93
Fife	8,100	2.24
Total	40,700	
Total 3 Areas	313,100	
Great Britain	640,900	

Source: Hall, 1987

Table 7.2 Great Britain: high-technology industry, employment changes, selected counties, 1975–81

	Employment 1981	*Employment change 1975–81*	
		Abs.	%
Greater London	91,448	−17,012	−15.7
Hertfordshire	45,060	5,914	15.1
Berkshire	19,732	7,600	62.6
Surrey	18,191	1,811	11.1
Hampshire	33,807	2,179	6.8
Greater Manchester	27,300	−2,980	−9.8
Lancashire	30,003	1,392	4.8
Strathclyde	23,575	−3,847	−14.0
Lothians	9,042	1,461	19.3
Great Britain	640,874	−42,429	−6.2

Source: Hall, 1987

of London. Inside these workshops, a series of leading engineers were truly the high-technology entrepreneurs of their day, virtually inventing the modern machine-tool industry before its subsequent development effectively transferred itself to the United States. The originator was Joseph Bramah, a Yorkshire farmer's boy, who came to London and took up cabinet-making; specializing in locks, he realized the need for machine tools. From his Pimlico shops came Henry Maudslay, whom he employed from 1789 to 1797, Joseph Clement, and other first-rate mechanics. Maudslay, born in 1771 in Woolwich, stayed with Bramah for eight years, but left to start his own shop, establishing himself with a partner in 1809 in Lambeth; his firm grew to employ several hundred men, making saw- and flour-mills, mint machinery and steam engines of all kinds, and training many other great names in British engineering, including Clement, Roberts, Whitworth, Nasmyth, Seaward, Muir, and Lewis.[1]

Out of these areas, by the end of the nineteenth century, came the beginnings of the British electrical industry. Clerkenwell and neighboring Hatton Garden, with its jewelry industry, provided precious metal contacts and contact by-metal to the infant electrical and radio industries; Ferranti in 1883, aged only 19, began his business here; in nearby Clerkenwell, Cossor established a small workshop in 1896, making a great variety of scientific glassware, including early fluorescent tubes, and the first British cathode ray tube (in 1902); by World War One the firm had turned to wireless transmitters and valves, staying with these products for the civilian market after 1918.[2]

At the turn of the century, firms began to move out from these cores in search of space; from Pimlico they went radially west to Hammersmith and

Acton, where they could draw on a concentration of engineering skills, and on large supplies of available female labor. Hammersmith had two electric lamp factories; during World War One these naturally diversified into radio valve manufacture, while new firms – such as the then small-scale business of Captain S.R. Mullard, working for Admiralty contracts – joined them in the same area.[3] Then, between the two world wars, the growth of both capital and consumer electrical goods caused huge expansion: between 1921 and 1951, employment in the electrical industries in London grew from 51,000 to 183,000, or three-and-a-half times. Firms moved out, and new firms joined them, mainly along radial lines from the existing concentrations. The west-London concentration spread out from Hammersmith and Acton into a wide sector embracing Park Royal, Hayes and Southall, the Great West Road and Slough. A study in 1933 found a total of 140,000 industrial workers in north and north-west London – of whom no less than 75,000 were in the Western Sector; 8,000 of these were in electrical equipment, and another 6,000 of them in gramophone manufacture, which during and after World War Two was to prove a particularly potent source of information technology spin-off.[4]

World War Two brought rapid conversion to military work coupled with dispersal from London for strategic reasons. Then, in the Cold War era of the 1950s, rearmament boosted demand for military electronics from firms that had diversified into the field during the war, coming directly out of the interwar radio, gramophone and lighting firms. They moved out in search of space. But by this time the Green Belt, established around the capital as part of the postwar regional planning package, proved a barrier, compelling firms to leapfrog into the neighboring counties to the west – Hampshire, Berkshire and Buckinghamshire, up to 60 miles (100 kilometers) away.[5]

But this could not be a complete explanation: many of the interwar electrical firms, growing from origins in Clerkenwell, had moved into north, not west, London; logically, therefore, there should have been an equal growth in that direction. There was another critical factor in the form of the Government (more accurately, Defence) Research Establishments, which had already located west of London before World War Two but had expanded during the war and again during the 1950s. The location of these establishments forms a particularly intriguing episode in economic history. Nearly all seem to have started as offshoots from older military establishments in traditional locations, and then to have grown incrementally in the same locations, or very near to them. The major naval establishments were (and are) located in and around Portsmouth, home of the Royal Navy since the sixteenth century, and in Weymouth; the Royal Aircraft Establishment in Farnborough, Hampshire, which was the most significant in terms of contracting, started in 1892 as a school of military ballooning and associated factory in next-door Aldershot, major base of the British Army. Another key establishment, the Atomic Energy Research Establishment, was established at Harwell in Oxfordshire during World War Two for security reasons.[6]

These steps might appear almost random in character. But they were not entirely so. As capital city, London was the center of the military and scientific establishments. The main bases of the army and navy, at Aldershot and Portsmouth, were located west of London for strategic reasons; indeed, Aldershot had been chosen through a deliberate search, in 1854. British top-level scientific research was highly concentrated in London, Oxford and Cambridge. Almost quixotically, the area contained a number of surplus royal palaces, which were pressed into service: the National Physical Laboratory was located at Bushy Park in 1900, the Radio Research Station in Ditton Park in the 1920s. And this in turn reflected the fact that this area west of London was the socially prestigious part of England, with long-standing associations with regal and aristocratic life; here for instance were Windsor Castle and Hampton Court Palace, and many other more minor palaces and country retreats. Sandhurst and Camberley, training centers for the British Army officer class, are for instance sited between Aldershot and Windsor.

The significance only became apparent with the growth of Britain's own military-industrial complex during the 1950s, when close day-to-day links developed between the research establishments and major electronics firms like English Electric, Edison Swan, General Electric, and Mullard, who tended to cluster around them and the Ministry of Supply (later, Ministry of Defence) headquarters in London. By the late 1970s, the South-East region had over 40 percent of all defense procurement expenditure, though official government policy was still to promote industrial development in other regions. Of the 57 UK-based contractors receiving £5 million or more for defense equipment in 1981–2, 43 were headquartered in the South-East. Defense contractors, interviewed for a research study, were quite clear about their need to be close to government agencies in the critical R&D and prototype stages.

In this sense, the London story points to the critical importance of traditional agglomeration economies in the maintenance of a defense-oriented innovative milieux. What happened, in the course of a century and a half, was that these agglomeration economies in effect relocated themselves, from tight artisanal quarters close to the seat of government in Westminster, to a slightly larger and looser complex which, nevertheless, was still extraordinarily locked into close contact with its governmental client.

During the 1970s, these agglomeration economies had a further effect: by building up a mass of skilled labor and subcontracting services, close to London and to Heathrow, they attracted inward investment by multinational companies, especially American ones, seeking a European base. Some major companies, like Digital, established a major presence in the area.

They located here not their peripheral, low-cost production activities – these went to other regions, like Scotland – but rather their national or even international base for R&D, management and marketing.

However, there was a real price to pay. Quite unlike the story in Silicon Valley, both larger contractors and smaller subcontractors in effect became locked into dependence on comfortable defense contracts. As American research has shown, the characteristics of such military contracting are so different from those of the commercial marketplace – indeed, contraposed to it – that they make it virtually impossible to adjust. Though the IT revolution of the 1970s and 1980s found some expression in Britain, with the emergence of entrepreneurial producers like Sir Clive Sinclair, it could not remotely compare with the explosion that was occurring in Silicon Valley. And this, without doubt, helps explain the extraordinary anomaly that, far from compensating for the decline in basic industrial employment, information technology has perversely played its share in the British decline: the growth of employment in London's Western Sector did not even keep pace with declines elsewhere, above all within London itself.

This points to the danger of creating a defense-based milieu. Apart from that, the experience of London demonstrates the difference between two styles of planning. At first sight, conscious planning did not have much to do with the growth of the Western Corridor. The westward drift of London's industry, which started the process in the 1930s, was totally unplanned; indeed, alarm at the process was one of the main triggers in developing regional policy after the war. But this policy, which was supposed to control industrial growth in the South-East and encourage it elsewhere, was simply ignored in allowing defense firms to locate and to grow here. More locally, policy at both national and county level was to restrain growth in the area around London, the new towns excepted. (One only of the original eight London new towns, Bracknell near Reading, is located in this sector.) There was a partial exception in 1970, when a major regional strategic plan identified the area bounded by Reading, Wokingham, Aldershot and Basingstoke as a major growth area; but the proposal met fierce local opposition, and more or less died.

The same conclusion goes for major transportation investment decisions. Though the area is often popularly called the M4 Corridor, for a freeway that runs through it, this was started only at the beginning of the 1960s and completed only in 1971; so it can hardly have been a factor. Heathrow airport is placed strategically between the corridor and London itself; but no thought of that entered into the original decision, taken during World War Two, to locate it here. Subsequently, its unequalled range of international connections was undoubtedly important in attracting inward migration of multinational firms, both from Europe and the United States, during the 1960s and 1970s. Reading was served by Britain's first high-speed train service, started in 1976; but this was in response to demand, not an attempt to encourage it. Thus, though these public policy decisions played a role in making the area more attractive to inward investment, this came about almost as a matter of accident.

At a slightly deeper level, though, key decisions played a role. The most important, as already seen, was the decision to place the defense research establishments west of London. And, somewhat perversely, tight local planning controls may have had a positive effect by creating a series of medium-sized urban centers, linked by very high-quality communications, against a very attractive green backcloth – thus maintaining a very high quality of life in the Corridor, far higher indeed than Silicon Valley achieved with a weaker growth control regime.

The conclusion, then, is that public policy played a major role, but more or less unintentionally: "if public policy is supposedly based on co-ordinated conscious decisions coupled with lively awareness of the *consequences* of the decisions made, the fact that the area has developed in the way that it has, with the speed that it has, is almost entirely accidental."[7] Decisions were taken in different agencies on narrow grounds, without awareness of their larger cumulative consequences. In particular, there was no apparent relationship to the development of regional policies either at the national level, which was aiming to divert dynamic industries out of the South-East altogether; or within the region, which sought to encourage the growth of new and expanded towns in other parts of the region. Yet the consequences did prove quite momentous, even though the resulting industrial district was insignificant in comparison with its American counterpart.

PARIS

While not universally heralded as a high-technology center, Paris in fact harbors the most significant concentration of technologically-advanced firms and public and private research laboratories in the whole of Europe. Most of them are found in the southwestern suburbs of the Paris metropolitan area, around the Plateau of Saclay, an area of 50 square kilometers (20 square miles) between the new town of St Quentin-en-Yvelines and the old town of Massy, bordered on the north by the valley of the Bièvre and on the south by the valley of the Yvette. The 15 communes in the area are the home of 102,000 people; about the same number work or study in the institutions and companies located there (Figure 7.2).

If any region deserves the label of "European Silicon Valley," at least in terms of the scale of research activity and high-technology industry concentrated in one location, this is it: in the late 1980s, within a radius of eight kilometers (five miles) of Massy were concentrated around 10,000 technologically advanced firms in sectors such as electronics, data processing, telecommunications, energy, and biotechnology.[8] And in the same area were

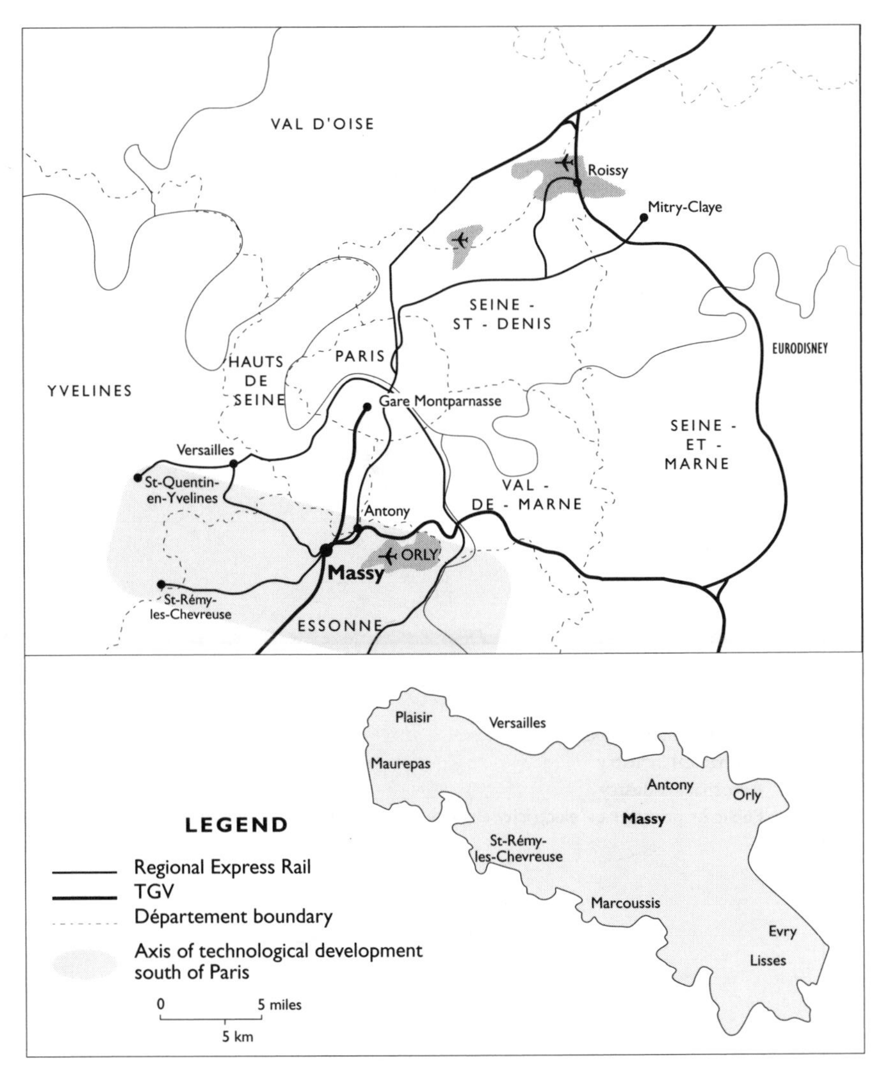

Figure 7.2 Paris: the Cité Scientifique Ile-de-France Sud and the Axe Sud
Source: Anon, 1989

found the University of Paris-XI, formerly the Faculty of Science (Orsay Campus) of the old University of Paris; 60 percent of the "*Grandes Ecoles*," the engineering and business professional schools that have traditionally constituted the elite of the French system; and some newer institutions such as the Ecole Polytechnique, Ecole Centrale, Institut Supérieur de Télécommunications, Ecole Nationale Supérieure de Techniques Avancées, Ecole Supérieure d'Optique, Hautes Etudes Commerciales, Hautes Etudes

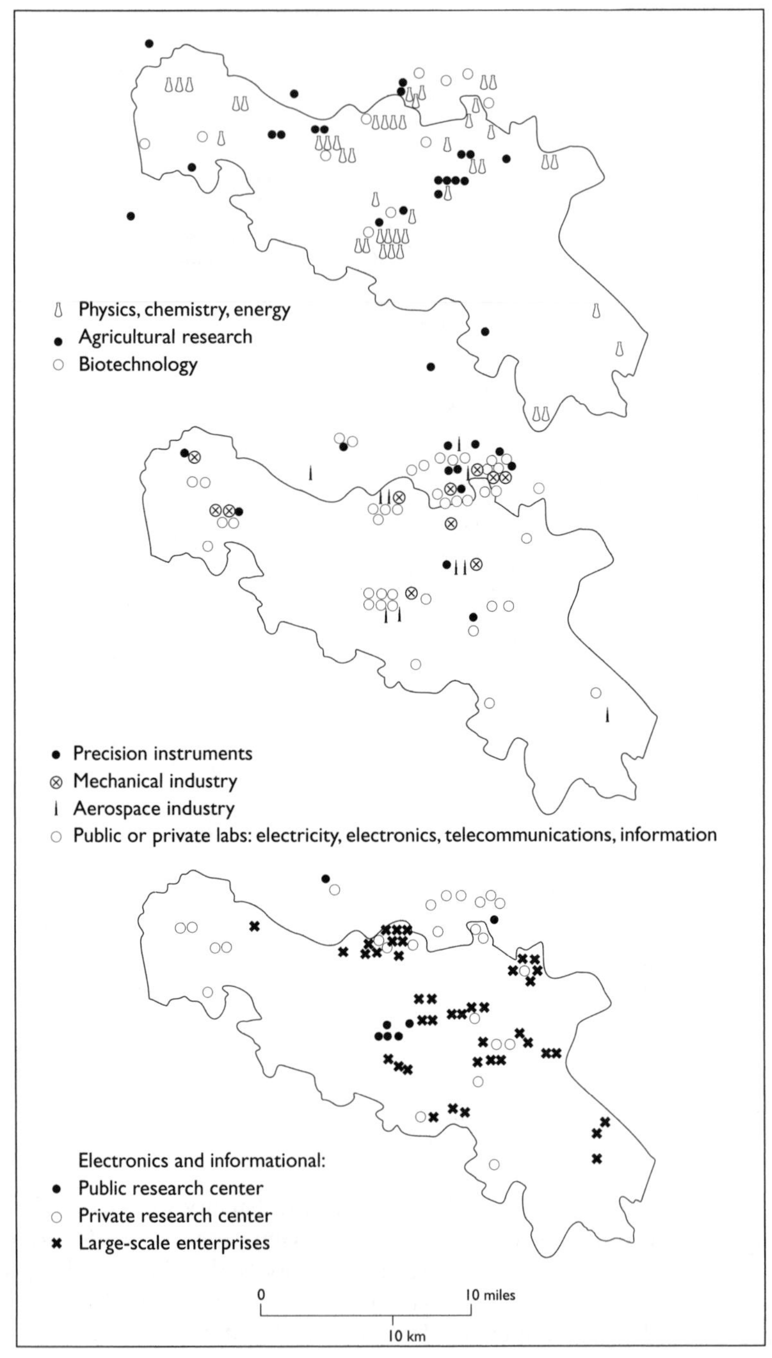

Figure 7.3 Paris Axe Sud: high-technology industries
Source: Decoster and Tabaries, 1986

Informatiques et de Gestion, and the Centre d'Etudes Supérieures et Industrielles. In the same area were found 43 percent of all French laboratories and institutes of public scientific research. In 1991, too, there were four privately-owned high-tech industrial parks, for selected innovative companies, in course of development. Altogether, the area contains about 35,000 researchers and 45,000 students (Figure 7.3).

These are impressive enough figures. But there is a deeper and ultimately more important question: to what extent can this impressive agglomeration be truly considered a high-technology complex or an industrial district, something that depends on the synergy developed within the area, and on the mechanisms of interaction between its various components? Clearly, all the ingredients are in place; the question is whether they are generating the necessary mechanisms of exchange. Before trying to address that question, though, we should seek to answer another: how did this high-technology agglomeration come into existence, in the shadow of one of the greatest and most prestigious cities in the world?

It was certainly not by design. True, in November 1983, the existence of this new technopole was institutionalized by formally constituting the "Association Cité Scientifique, Parcs et Technopoles en Ile-de-France Sud," which joined together local communes, *Sociétés d'Economie Mixte*, new town authorities, regional and departmental bureaux of infrastructure, and independent research institutes; all under the auspices of the Ministry of Industry and Research, and with the blessing of President Mitterrand. But this was a typical French theatrical gesture, giving a top-level state imprimatur to something already established as a fact, in a process that had been going on for at least 20 years. Yet, such public recognition is not and will not be inconsequential. As a result of the new visibility of the Paris-Sud technopole, the draft for the new regional Plan of Paris (*Livre Blanc*) of 1990 designates the Massy-Palaiseau area as a regional development pole supported by the two new towns on its borders, St Quentin-en-Yvelines and Melun-Senart.[9] The proposal is organized around the concept of exchange functions between research, industry, services, and the marketplace, and includes the development of office space, a cultural and scientific center, a high-level service center, and a permanent research market, articulated in an "urban regional center with a European dimension" that will include a TGV (*Train à Grande Vitesse*) station, two RER (Regional Express Rail) stations, a bus station, two regional parks, a conference and communications hall, a scientific education hall, a music institute, a university institute, seven hotels, cinemas, commercial facilities, and sport installations. The communications center, organized around the TGV station, has been opened in 1991.

What then were the origins of the various components of this technopole and the factors that aided their development, as well as their location in this particular area? Paris-Sud seems to result from a combination of three factors, originally independent, that came to reinforce each other: the location of the government nuclear and defense oriented research centers in the 1950s, followed by the decentralization of part of the Faculty of Sciences of the University of Paris; the location pattern of high-technology manufacturing industries in Paris from the early 1960s onward; and the impulse given to the region by the 1965 *Schéma Directeur de la Région Parisienne*, operationally translated into the construction of two nearby new towns, the reinforcement of the transportation axes towards the South, and the government-initiated location of several *Grandes Ecoles*.

The origin of the entire process, the location of the government research centers, was marked by a strange combination of accidental factors and of the perceived image of the area. There were large tracts of vacant land that had become government property after their confiscation from wartime German collaborators. In 1951, Frédéric Joliot-Curie chose one of these tracts as the site for the new Nuclear Research Center, which would become the brain of the French nuclear program. Simultaneously, a nearby site was chosen for the Jet Propulsion Test Center. Historically, the area was highly symbolic in terms of social prestige, since its main urban center, a few kilometers to the west, was the very symbol of royal power and French historical glory, Versailles. And the nearby location of the French Army Elite School, at St Cyr-l'Ecole, added emphasis to the aristocratic functions of the area.[10] A little farther south, the establishment of the Faculty of Sciences at Orsay, in 1954, was the first major attempt to decentralize the overcrowded Sorbonne, using the unique transportation facility of the Ligne de Sceaux, an old suburban railway that connected these southern suburbs directly to the core of the Quartier Latin and the Sorbonne.

The outward movement of industry was the decisive factor in fostering the growth of the future technopole. It was well under way by the early 1960s, as shown by Castells's study of industrial location patterns in the metropolitan area of Paris.[11] Analyzing all firms seeking to establish a new location in the Paris region between 1962 and 1965, a total of 930 cases, Castells found a distinctive location pattern for high-technology, R&D-based firms: they placed a high value on location in socially-prestigious localities within the metropolitan area of Paris. Given the lack of suitable industrial space in the extremely dense city of Paris, the area that could provide the right combination of low-density, available land, and higher social status was the southwest quadrant of the region. Because of the early construction of two radial freeways to the south and west of the region, relocation was relatively easy. However, Castells's statistical analysis showed that for these companies such functional considerations would have allowed them to settle in almost any sector, so long as they could stay in the Paris metropolitan area; efficiency was less important than prestige and social status.

Thus, Castells's analysis of the selected location pattern for innovative-type firms uncannily predicts – even at that early date – the site of the future technopole of the 1980s. A critical location decision was that of Thomson-CSF, which located its major research center at Orsay. The proximity of a secondary airport at Vélizy-Villacoublay, often used for flight tests, was another factor in anchoring in the area research facilities and auxiliary firms of the aircraft industry. These firms had frequent contacts with the factories of the Dassault aircraft company, located in the industrial town of Courbevoie in the northwest sector of the Paris region. Thus, a western location for these companies was by no means eccentric to their main focus of activity; yet they benefited from a better environment and higher social status. In all, from the 1960s onward the southwest sector of Paris began to become a magnet for high-level research centers in nuclear energy, electronics, and aviation: the core of French high-technology industry, directly linked to the Government's priority programs in defense and nuclear energy, the two pillars of the Gaullist dream of a powerful, independent France.

The third key element contributing to the formation of the Paris-Sud technopole was the French planning process itself, and its direct influence on the decisions to locate (or relocate) establishments of research and higher education. The 1965 *Schéma Directeur*, directly inspired by de Gaulle through his personal appointee Paul Delouvrier (in a partnership often likened to that of Napoléon III and Haussmann), will remain as one of the most ambitious master plans in the entire history of planning, as a basis to guide metropolitan growth and set up the spatial basis of economic and political power. The underlying philosophy was unambiguous: the future Europe would have its center in France, and France would (as always) have its center in Paris; thus, the metropolis should be ready for it. Projecting a 14–16 million population for the whole of the "Paris District" (roughly equivalent to the present-day Ile-de-France region) by the end of the century, the *Schéma Directeur* went on to organize and support such growth by developing transportation infrastructure, creating new directional centers in and around Paris, and building the new towns that would ensure a dynamic process of intra-metropolitan decentralization. The *Schéma Directeur* envisioned the future importance of high-technology industrial and research activities and, partly based on the research programs conducted under its auspices, explicitly favored their concentration in the southwestern suburbs. In a symbolic move, Serge Goldberg, director of the official institute that prepared the *Schéma* (Institut d'Aménagement et d'Urbanisme de la Région Parisienne, IAURP), left to take up the job of director of the new town of St Quentin-en-Yvelines, the basic urban center to support the high-technology growth area.

In the late 1960s, when the Ides of May started to project its somber clouds over de Gaulle's grandiose dreams, the Government engaged in an ambitious plan for the decentralization of key *Grandes Ecoles* and public research centers into the area, as the most direct way to define the character of the area and to enhance even more its social status. Throughout the 1970s, in spite of the disintegration of the Gaullist empire, the impact of the farsighted decisions taken to implement the *Schéma Directeur* continued to be felt. The new towns were constructed, and helped to provide housing, facilities, and office space for the new research institutes and industrial firms. The *Grandes Ecoles* did follow the Government's request, if only to prevent a much bigger threat to force them out of the Paris area. And the dramatic expansion of the Parisian transportation and telecommunications network closely linked the area to the core of Paris, while keeping the necessary distance to develop high-technology activities in a relatively autonomous, self-contained space.

Thus, a unique combination of factors – purely accidental forces, the spontaneous locational pattern of high-technology industries, and the impulse of Gaullist governments to implement the industrializing vision of the *Schéma Directeur* for Paris – contributed to the formation of Europe's most prominent *de facto* technopole. But the critical question remains: to what degree has the entire Paris-Sud complex, created in this rather extraordinary way, become a genuine milieu of innovation?

Here, academic studies tend to reach a negative conclusion: this is not a true technopole, at least not yet. It lacks the necessary synergy between science and industry; or, indeed, between large firms and small firms, or among large firms themselves.[12] Thus, Decoster and Tabaries conclude that industrial–research connections are too weak to be able to speak of a true growth pole. Similar conclusions are reached by another expert, Jean-Claude Perrin (1988: 155); he considers that synergy and cross-fertilization in the area are minimal, fundamentally for problems of communication: "Each unit behaves like a citadel jealous of its power and of its supposed authority."

In fact, it seems as if the different components of this high-technology agglomeration never truly merged. Large high-technology firms in France tend to follow a pattern of aloofness *vis-à-vis* their immediate environment – a pattern we also found at Sophia-Antipolis (Chapter 5); they are more likely to engage in international alliances within a global strategic framework than in intra-regional linkages. Each large firm tends also to have a network of ancillary firms that primarily supply the parent firm, thus making it more difficult to develop horizontal industrial contacts and technological cross-fertilization. Traditional government research centers, such as the Saclay Nuclear Center, have been traditionally secluded, for two reasons: the proud isolation of the French public research system, which traditionally avoids contact with the private sector and even with the university system; and the tight security procedures linked to the military or para-military character of many of the important programs. As for the core of the higher education institutions in the area, the *Grandes Ecoles*, their culture was historically built upon the social and scientific superiority of the elite corps they create and reproduce, thus aiming to irradiate their power and influence throughout the whole of the French economy and society, and precluding any parochial preference for neighboring firms or institutions.

Indeed, the 1965 *Schéma Directeur* itself never thought of a high-technology complex in the sense we understand in the 1990s, built on synergy and networking. It simply reserved a traditionally aristocratic space for the aristocratic function of leading the whole of Paris, and thus the whole of France, into the new age of international business and science-based industry. But it was never intended that the elements of this *cité scientifique* would fuse into some new form. Their destiny was simply to be the inhabitants of this "City without Walls," in Mitterrand's words. It was without walls because it was intended to be universally projected, on the basis of the excellence of each one of its components, without ever understanding that modern science and technology require the invisible walls of organizational and commercial networks to close the milieu on to itself to generate the necessary synergy within, before it can project itself into the world as a collective unit of innovation.

Nevertheless, the very existence of the Paris-Sud high-technology agglomeration has triggered a process of economic and technological dynamism, that has gone far beyond the intentions of any one of its unwitting founders. The exceptional concentration of firms has permitted the development of numerous spin-offs from large companies, to the point that the southern suburbs have the highest ratio of small business firms in the entire Paris metropolitan area. Decoster's analysis documented the exceptional record of new-firm growth in this area in the 1976–83 period, particularly in the leading edge industrial sectors.[13] On the other hand, perceiving the need to articulate this mammoth high-technology concentration to the wider French economy, the Government moved in 1984 to establish several Regional Centers of Technology Transfer, which would diffuse the research findings from the *cité scientifique* within the industrial structure of the country. And finally, the 1990 *Livre Blanc*, aiming at a new projection of Parisian glory well into the twenty-first century, transforms the *cité scientifique* into a directional, high-technology center forming the core of the new Parisian industrial base, at what is intended to become the transportation crossroads of the new Europe.

Paris-Sud may thus not be a technopole in the archetypical sense, but it does propel Paris into the twenty-first century's industrial world. Yet it does so with an additional and somewhat perverse dimension: its profound originality is to bring together – in one territorial space – public research, private companies, and elite schools of the highest level of excellence, while creating every possibility for them to continue studiously to ignore each others' existence. And, final oddity, all this is done with great publicity, under a public-relations umbrella formed by cleverly-labeled government bureaucracies. It is *magnifique*, but it may have little relevance to the coming twenty-first-century *guerre de l'innovation*.

TOKYO

The Keihin region of Japan, embracing metropolitan Tokyo and its surrounding prefectures of Kanagawa, Saitama and Chiba, is without doubt the leading high-technology industrial area on the entire globe. And, in sharp contrast to its leading rival in Southern California, its primacy is based firmly on production for the consumer market, so that it is insulated from the vagaries of international conflict. Indeed, with just under one-quarter of Japan's manufacturing workforce and just over one-quarter of total manufacturing output, it is the largest industrial region in Japan and one of the largest in the world.[14]

Doubtless because it is so consumer-oriented, in comparison with almost all other such metropolitan areas, it has shown an astonishing rise in productive capacity in the half-century since the end of World War Two, when its factories lay in ruins. But that should not be allowed to obscure the fact that Tokyo, though not as old a manufacturing city as London or Paris, has nevertheless an industrial history that goes back for well over a century.

Within this industrial heartland there is an inner core, embracing the southern and eastern sides of Tokyo, next to Tokyo Bay, and the neighboring Kanagawa prefecture. With 1.7 million manufacturing workers in some 120,000 establishments, in 1980 Tokyo and Kanagawa together accounted for 16.4 percent of all manufacturing establishments in Japan, 15.7 percent of employment and 18.7 percent of value of shipments (Table 7.3). For the electrical machinery industry, the share was even more striking: with 332,000 employees in more than 10,000 separate establishments, the area accounted for 29.6 percent of Japan's establishments, 24.5 percent of employment and 28.7 percent of value of shipments (Table 7.4). But most notable of all was the performance in electronics, where the area accounted for 56.2 percent of establishments and 78.8 percent of shipments of computers, and 35.3 percent of establishments and 58.6 percent of shipments of automatic switching equipment (Table 7.5). The wider Keihin region, in 1982, had 68 percent of the national total of establishments in optical communications equipment; 62 percent in space apparatus; 58 percent in computers; and 47 per cent in industrial robots. It also had no less than 50 percent of private R&D laboratories in engineering, and 37 percent of researchers working in national engineering institutes, and 30 percent of science and engineering university faculties. Particularly significant, perhaps, no less than 60 percent of Japanese information-processing engineers were found here (Figure 7.4).[15]

Tokyo-Kanagawa is thus the heart of the modern Japanese electronic miracle. And, in recent years, as industry has decentralized out of Tokyo, Kanagawa has become the manufacturing core of the region. Among the 47 prefectures in Japan, Kanagawa ranks ninth in number of factories, second in value of manufactured goods, and first in value added per factory as well

Table 7.3 Manufacturing in Tokyo and Kanagawa as percentage of national total, 1980

Area	*Establishments (%)*	*Employees (%)*	*Value of shipments (%)*
Tokyo	13.2	9.4	9.9
23-Ward Area	12.0	7.4	7.7
Kanagawa	3.2	6.3	8.8
Kawasaki	0.7	1.4	2.4
Yokohama	1.1	1.8	2.3

Source: Ministry of International Trade and Industry, *Census of Manufactures 1980*

Table 7.4 The electric machinery industry in Tokyo and Kanawaga as percentage of national total, 1980

Area	*Establishments (%)*	*Employees (%)*	*Value of shipments (%)*
Tokyo	20.7	12.1	13.5
23-Ward Area	15.8	6.5	6.9
Kanagawa	8.9	12.4	15.2
Kawasaki	3.1	4.0	4.7
Yokohama	3.0	3.5	4.2

Source: Ministry of International Trade and Industry, *Census of Manufactures 1980*

Table 7.5 Electronics-based industries in Tokyo and Kanagawa as percentage of national total, 1980

Area	*Establishments (%)*	*Value of shipments (%)*
	Digital computers	
Tokyo	28.1	24.9
Kanagawa	28.1	53.9
	Automatic switching equipment	
Tokyo	19.6	11.0
Kanagawa	15.7	47.6
	Integrated circuits	
Tokyo	5.1	12.9
Kanagawa	11.2	21.9

Source: Ministry of International Trade and Industry, *Census of Manufactures 1980*

as per employee.[16] Its gross domestic production quadrupled from ¥4,770 billion in 1970 to ¥19,005 billion in 1985: an average annual growth rate in real terms of 3.5 percent.[17] During that time, with a brief exception during the oil crises of the mid-1970s, Kanagawa has recorded about six percent of the Japanese Gross Domestic Product (GDP).[18]

During the period since World War Two, the era of Japan's dramatic rise to industrial preeminence, the country's key manufacturing industries have changed from energy-intensive ones to knowledge-based ones – a change that

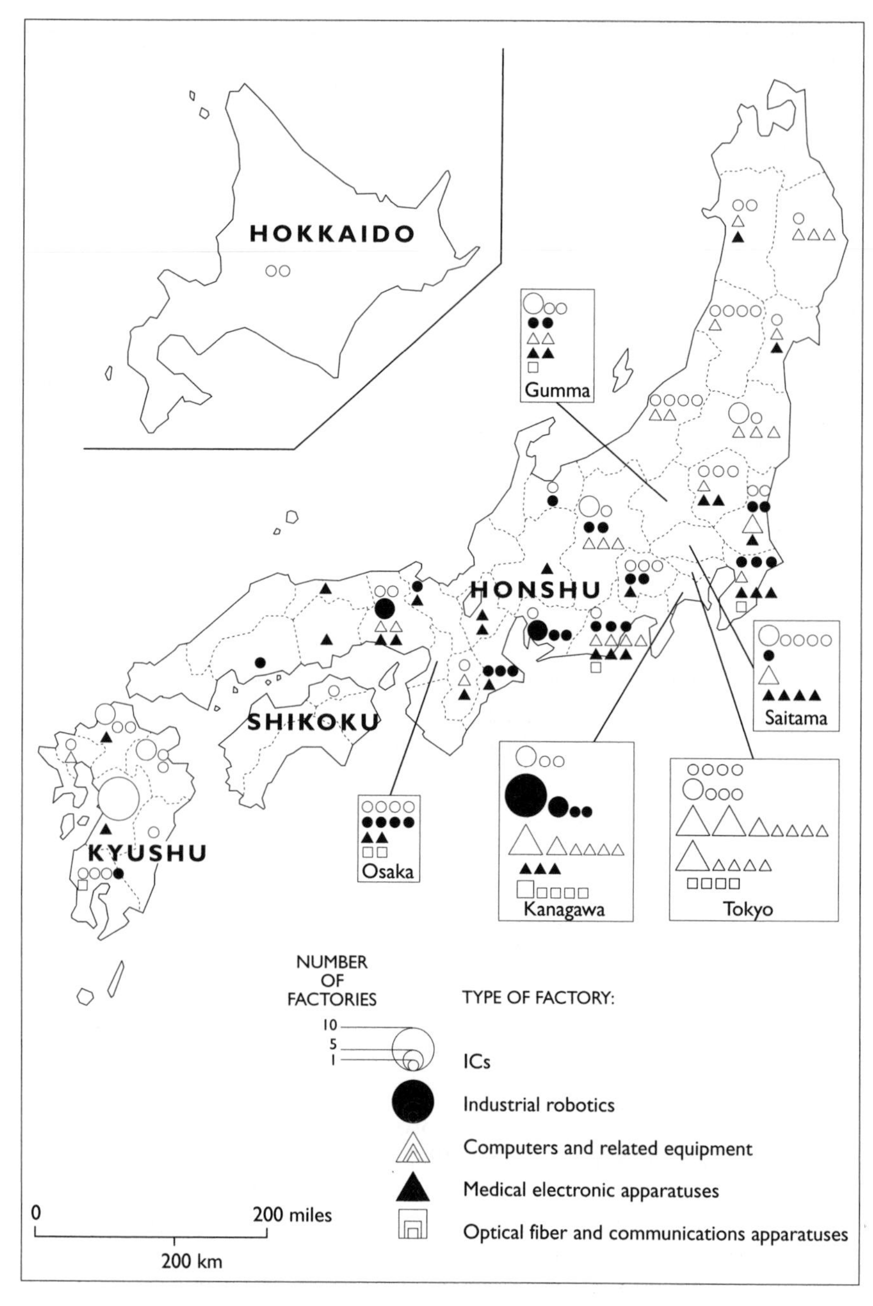

Figure 7.4 Japan: location of high-technology factories
Source: Nishioka, 1985

is very clear in the Keihin region, and particularly so in Kawasaki and Yokohama. Here, the original industrial development was in energy-intensive industries, such as steel and chemicals, which started to develop on reclaimed land in the coastal areas of Kawasaki and Yokohama from the 1910s, followed by the creation of petrochemical complexes half a century later. But, since the 1960s, the region's leading industries have become knowledge-based ones, the most notable being the shift from electrical engineering to electronics.

A crucial feature of this industrial complex, since at least the 1930s, has been the intimate relationship between very large corporations – now, among the best-known household names in the world – and a host of small subcontractors. The assembly factories of the large corporations, which recently tend to have moved to the outskirts and peripheral areas, depend on the supply of sophisticated parts from those small and medium-sized factories, which are still overwhelmingly located in Tokyo itself. Thus, Tokyo remains a seedbed for the entire Japanese engineering industry, including its high-technology leading edge; this small-scale activity is closely connected with the concentration of R&D and information functions there. In the Akihabara area, an inner-city district known to all Tokyo visitors as the capital's "electronics city," more than 100 independent system firms are concentrated, creating new information on electronics-related hardware and software.[19]

Some history

The reasons for the industrial dominance of the Keihin region, and especially Kanagawa, go back to the Meiji restoration of 1868, and indeed even before that.[20] They begin with the removal of the capital from Kyoto to Tokyo, then known as Edo, under the Tokugawa shogunate in the seventeenth century. This greatly increased traffic along the main east–west road (the Tokaido) between Edo and the old capital region of Kansai (Kyoto-Osaka). On this, and on four other roads radiating from Tokyo, post stations were established to provide lodging for travelers.[21] The towns, post stations, and markets of the region had originated in the castle towns established by feudal lords of the Sengoku period (the Period of Warring States), as did the transportation system they built up, and the trade and artisanal manufacture they patronized and protected within their domains. Odawara was one of the most prosperous castle towns; Uraga was a busy port that controlled access to Edo.[22]

These towns served purely internal trade, since in this period the country was effectively closed. But all this ended with the arrival of Commodore Matthew Perry of the United States East India Squadron in 1853, and his return the following year. He established his base in the Yokohama area; the resulting Treaty of Kanagawa, ending Japan's long years of isolation, was signed here; then, following the first US–Japan commercial treaty of 1858, the shogunate began to develop the former fishing village as an international trading port with a new, planned city. Yokohama's trade boomed; foreigners settled there and established trading houses, soon joined by Japanese merchants from Edo.

Innovative financial magnates emerged and embarked on innovative projects, such as land reclamation in Tsurumi. New financial institutions, such as the Bank of Japan and the Yokohama Specie Bank, were founded there or in Tokyo. By 1893, the port had a colony of 5,000 foreigners.[23] Yokohama became a base for exporting raw silk, followed by other export-related industries; logically it became a node of transportation, and the point where foreign technologies were introduced. The first telegraph line was established here as early as 1869, and by 1872 a railway connected the port with Tokyo, which became the political and economic center of the country after the Meiji Restoration. Soon, the old Tokaido road was paralleled by the construction of a railway, the Tokaido Line.

These were clear natural advantages; but they were resolutely acted upon. In 1911, the Yokohama city council exempted factories from municipal taxation. This, coupled with port improvements and investments in railways and electric power, attracted industrialists from Tokyo.[24] A factory of the Tokyo Electric Company (now Toshiba), which utilized electric power generated by the newly-opened Daishi Electric Railway, was opened at Kawasaki in 1913, with financial and technical support from the American General Electric company.

Thus Kanagawa attracted industry, especially large factories, faster than any other prefecture near Tokyo. The huge Tokyo market attracted not only producer-goods heavy industries such as shipbuilding and steel, but also consumer-goods industries like food and drink, home electric products, records, and automobiles. The heavier, bigger plants located on reclaimed coastal land, close to the port; lighter ones were logically displaced inland into the Kanto Plain, the largest area of flat land in Japan.[25]

Yet despite all this activity, for a very long time – until immediately before World War Two, in fact – Japan's main center of industrial activities was not the Keihin region but the Hanshin (Osaka-Kobe) region. At the first modern statistical count of 1874, the leading industrial prefecture was Kyoto, followed by Osaka; Tokyo was ninth.[26] What helped the Keihin region was government support. The Meiji Restoration aimed at the rapid introduction of modern industry, and government model factories were established in various places, but above all in Tokyo; most were later transferred to private enterprise, joining private companies already established.

The region's engineering industry was first established early in the Meiji Era in the Ginza-Kyobashi district, which is now the commercial center of Tokyo, and then progressively expanded to the south and east through the reclamation of low-lying land, on the shores of minor rivers flowing into Tokyo Bay and the bay shore itself, especially the coastal plain of the River Sumida, at the fringe of the city. Engineering factories were located in the whole inland part of Kawasaki and Yokohama, and much later – in the 1960s – the core of the industry would consolidate there. The region thus assumed its original shape by 1910, gathered momentum in the recovery from the *Showa* recession in the early 1930s, and finally came to surpass the Hanshin region by 1940, immediately prior to Japanese involvement in World War Two.[27]

But in doing so, its structure evolved. Before World War One the area was dominated by small workshops. But first in the period after the Sino–Japanese War of 1905, and then more sharply in the 1920s and 1930s, there was a noticeable trend toward the mechanization of traditional industries, the penetration of large-scale capital, and the diversification of manufacturing.[28] The electrical industry, which had developed in the late nineteenth century through small entrepreneurial firms, such as Tanaka Engineering Works, Hakunetsu-sha, and Meiko-sha, as well as the foreign-affiliated Nippon Electric Corporation and *zaibatsu*-affiliated Fuji Electric and Mitsubishi Electric, came to substitute for imports between the two world wars, with the emergence of large companies including Tokyo Shibaura Electric, Hitachi, Mitsubishi Electric, Nippon Electric Corporation, and Fujitsu (a subsidiary of Fuji Electric). Over 50 years later, these major names in electrical engineering-based industry are the leaders of electronics-based industry in the world as well as in Japan, though their transformations have not always been smooth.

For example, Toshiba was founded in 1939 by merging a firm of heavy electric apparatus – Shibaura Engineering Works, which went back to 1875 and which had diversified from general to electrical engineering after 1900 – and a firm of consumer electric appliances, Tokyo Electric Corporation, which started in 1900 and diversified from light bulbs into radio and home appliances. It continued to rely on both sectors as profit centers. In the postwar era the company continued to diversify in response both to technological opportunities and growing demands from an increasingly affluent market, into washing machines, rice cookers, radios and TV sets in the 1950s, into refrigerators and air-conditioners and color television in the 1960s. Making a start in computers as early as the 1950s, it took a bold decision to exit from mainframe computing in 1978 and then, from 1983, concentrated on laptops – a move, based on American market research, that proved extraordinarily perspicacious. Industrial electronics, however, came to account for over half of Toshiba's sales in the 1980s.

Fujitsu was, by contrast, founded by Fuji Electric as a subsidiary specializing in communications equipment. Now, Fujitsu ranks first in the Japanese computer market. Nippon Electric Corporation (NEC) was established by Western Electric as its affiliated company specializing in communications equipment. With its strategy of "C&C" (computers and communications), NEC is now a leading company in computers and semiconductor devices as well as in its original specialty.[29]

These three companies, Toshiba, Fujitsu, and NEC, have been based in the Keihin region throughout their histories. Toshiba has over 10 factories in Kanagawa. Fujitsu's central factory is located in Kawasaki. NEC has factories in Kawasaki and Yokohama. Each company has its R&D center in Kawasaki. All the companies are also headquartered in Tokyo. Through the continued presence in the Keihin region, the three companies have contributed to the transformation of the region's electric machinery industry.

After the 1923 earthquake, small-scale industry gradually disappeared, and the industrial belt centered on Kawasaki and Tsurumi came to be dominated by major corporate complexes already located in the area, by the various factories that moved from Tokyo and Yokohama in the wake of the earthquake, and by new industries such as the automobile industry that had recently sprung up in the area.[30] Tokyo Electric had already opened a factory at Kawasaki as early as 1913. In 1919, Toyo Electric, which was specialized in electrical equipment for railways, constructed a factory in Yokohama. Shibaura Engineering Works, later to merge to form Toshiba, purchased a section of reclaimed land at Tsurumi in 1922 and moved both its headquarters and production operations here from Tokyo in 1925.[31] Shibaura specialized in the manufacture of electrical equipment, heavy electrical machinery in particular; while Toshiba's other half, Tokyo Electric, in addition to manufacturing light bulbs, various types of lamps, and thermometers, also produced a wide range of consumer electric goods.[32] In 1925 Furukawa Electric, which had founded Fuji Electric as a joint venture with the German firm Siemens, built a huge 158,000 square meter (189,000 square yard) factory at Kawasaki for the manufacture of electric machinery and equipment.

These major companies, which emerged in the region between the two world wars, would later transform themselves into world leaders of the electronics industry. But their astonishing success in that regard was underpinned by certain distinctive features of their industrial organization, which already began to emerge in the 1930s. One is cooperative subcontracting, which allows them to respond flexibly to rapidly-changing markets. A prototype can be traced back to the Government's program for wartime production during the period 1931–45. Later, large corporations made efforts to convert many subcontractors into contract assemblers and systems components manufacturers during the high-growth period in the 1960s, in order to keep up with rapid growth of markets, and to increase their product range. Subsequently, in the 1970s and 1980s, many subcontractors came to possess high levels of technology and skill, with cooperative subcontracting practices.

By 1920, Kanagawa prefecture was already among the top prefectures in terms of manufacturing production; by 1930 it was second. Then, after the country's entry into the "Fifteen-Year War" against China in 1931, military production rapidly increased. Construction of private factories, under the control of the army and navy, expanded the industrial belt to inland areas and along the Tokaido Line to the west of Yokohama.[33] Particularly notable was the growth of the electrical industry: the number of factories grew from only 8 in 1926 to 20 in 1935 and no less than 141 by 1942, reflecting the impact of war production, with a particularly strong increase in heavy electrical equipment and telecommunications equipment. The new factories were predominantly located in the inland areas of the prefecture, away from the coast.[34]

Devastated at the end of World War Two, the Keihin industrial belt was in a parlous condition. In only 10 months during 1944–5, 22,885 tons of bombs were dropped on the Tokyo-Kawasaki-Yokohama area, reducing the area to rubble for the second time in less than a quarter century.[35] In addition to the damage, the failure to maintain and renew plant and equipment during the long war years contributed to the chaotic state of the factories in the region. Firms adapted as best they could. Toshiba, based in Kawasaki, cut its wartime staff of over 100,000 workers by more than half and produced a variety of products ranging from heavy electrical machinery for civilian industry to light bulbs, radios, washing machines, vacuum cleaners, and electric heaters ordered by Occupation forces, soon expanding production for the Japanese domestic market. Fuji Electric, whose Kawasaki plant had been heavily damaged by the air raids, reduced its wartime workforce of more than 15,000 to less than half, and started over again as a producer of medium-sized electrical motors, then diversifying into the production of multi-purpose motors, adding machines, and electric fans, which had not previously been its major lines of products, as well as new products such as agricultural machinery, electric heaters, and other smaller appliances.[36]

However, as the national economy recovered and entered into a period of high growth, the area rapidly recovered. In the process, from the mid-1950s onward it became the main arena for the development of huge industrial complexes (*keiretsu*) controlled by banks;[37] and the area became increasingly dominated by the engineering industry, with large firms putting work out to a great number of small subcontracting establishments.[38] By 1959, the Keihin industrial belt accounted for 18 percent of all manufacturing establishments in Japan, over 21 percent of employment, and over 24 percent of value of output. And by now, there was a clear distinction between Kanagawa, where over 40 percent of employees were in giant establishments with more than 1,000 workers, and Tokyo, where over 70 percent were in establishments with less than 300.[39] The large Kanagawa firms were chiefly in coastal Yokohama and Kawasaki; in the Shonan area and the inland areas of Yokohama and Kawasaki, west of the Tokaido Line, the majority of engineering factories were medium-sized.[40]

The process of decentralization

As the postwar boom continued, huge land reclamations took place in Kanagawa's coastal industrial belt for heavy, space-consuming basic industry.[41] Other industry migrated to inland areas, where industrial parks attracted small and medium-sized firms, including machinery companies which subcontracted to the larger factories located in the coastal industrial belt.[42] Factories from Tokyo moved to these same inland areas of Kanagawa, outside the four cities of Yokohama, Kawasaki, Yokosuka, and Miura.[43] Thus the belt expanded southwestwards into previously rural areas, such as Honmoku, Isogo, and Oihama.[44]

By the 1960s and early 1970s Tokyo, Kawasaki and Yokohama were recording declines in manufacturing employment, though engineering and electrical employment continued to grow in Tokyo, much of it still in small factories that subcontracted to the giant firms farther out.[45] Employment increased to the north and east of Tokyo – at Urawa in Saitama prefecture and Chiba in Chiba prefecture – and generally in inland cities around the fringe of the Tokyo metropolitan area.[46] Between 1975 and 1986, when manufacturing establishments in Kanagawa increased by no less than 27.6 percent, the increase was particularly strong in the outlying Ken'o and Shonan areas, indicating that the manufacturing center of gravity was gradually shifting from the eastern coastal area to inland areas. Electrical machinery factories rapidly increased, particularly in the Ken'o area where new industrial parks attracted high-technology firms.[47]

This reflected significant underlying trends. With the shift to high-value-added, low-weight products, the attraction of the coastal areas weakened; microelectronics-based automation reduced labor inputs, making it easier to locate outside the urban labor pools; rising land prices in central areas forced firms to seek cheaper locations; labor supplies were more readily available in the outlying areas, as workers migrated outwards; better road and communications networks improved access to the inland areas; finally, construction of large factories was prohibited in Yokohama and Kawasaki, making it difficult to expand production there.[48]

Thus, though the core of the Keihin region extends outwards for some 50–60 kilometers (30–40 miles), from the 1960s the machinery industry was diffusing to its very fringe. At the northern periphery the precision machinery industry developed, and in the south of Saitama prefecture (west of Tokyo) a machinery industry grew from the casting industry. Factories moved from the southern part of the region, fanning out to cover the entire fringe from the southern outskirts to the western outskirts, and small concentrations grew there, initially in the 1960s. Next, from about 1965, this development extended to the northern outskirts, especially along the main national roads.[49]

Down to 1970, this process of decentralization was an internal one, within the Keihin region. But from 1960 the region's growth gradually slackened, turning into an absolute decrease from 1970 onward. From that date, dispersal of the factories of the large enterprises beyond Keihin – in large measure to escape locational controls in the built-up areas – became general, and the region's relative position declined.[50]

There are however limits to this process. Factories that disperse outside the Keihin region are generally located within an eight-hour trip by night truck service from the inner Keihin region.[51] In all such peripheral districts, the density of manufacturing firms is still low, but leading enterprises – motor vehicle factories of Nissan, Isuzu, Hino, and Honda, and electrical factories of Toshiba, Hitachi, Sony, and Nippon Electricity – have located themselves here.[52] A large assembly factory for Canon cameras is now located in Fukushima City, 300 kilometers (180 miles) northeast of Tokyo, though all the subcontracting parts factories are still in Tokyo.[53]

The majority of the small and medium-sized producers, who subcontract for the large companies, are still concentrated in the southern part of the Keihin region, particularly in southern Tokyo. For example, 55 percent of the subcontractors of 14 electric machinery factories, and 85 percent of the total of 14 machine tool factories are in this southern district.[54] And the Keihin region retains its position as Japan's manufacturing center, which it has maintained ever since the 1930s – even though the single largest prefecture, in terms of value of manufacturing shipments, is now Aichi prefecture, based on the city of Nagoya.

Industrial structure: mega-corporations and mini-subcontractors

The resulting structure of industry in the region is dominated by large firms, above all in Kanagawa. Establishments with over 300 employees, which accounted for only 1.3 percent of the total establishments in Kanagawa, had 50.1 percent of the total employees, and produced 67.8 percent of the total value of output.[55] Most of these large Kanagawa firms are headquartered in Tokyo; a 1959 survey showed that the average Tokyo-based firm was more than ten times as large as the average local firm.[56] The gap was largest in the machinery industry, where firms with less than 300 employees employed only 25 workers on average, while ones with over 300 employees employed an average of 2,053 persons.[57]

Many of these large Kanagawa factories – in a 1958 survey, about half – relied on subcontracting, obtaining some 11 percent of their value of output from them. Many of these subcontractors, especially the more specialized and skilled ones, were located in Tokyo; 41 percent of the Tokyo contractors were cited as offering "special techniques," as against only 28 percent for local contractors. The reason appears to lie deep in history: in Tokyo, subcontracting to small and medium-sized producers developed early in the process of industrialization, from the late nineteenth century onward, while Yokohama is a relatively new industrial area dominated by large producers, which did not develop the same linkages to local small and medium-sized producers.[58] Perhaps surprisingly, Kobayashi suggests that by the 1980s linkages between large corporations and the small and medium-sized firms in Kanagawa had become weaker, even though during the 1970s the latter had improved their technology and management.

There is however no doubt that one of the sources of innovativeness of the electrical machinery industry in the region is the dense networks of small and medium-sized firms in Tokyo – which, as observed in the decreasing size of establishments in Kanagawa, now seem to be extending into inland areas within that prefecture. The assembly factories of large corporations located on the outskirts of the wider Kanto region – comprising Tokyo, Kanagawa, Chiba, Saitama, Ibaraki, Tochigi, and Gunma prefectures – depend on the supply of sophisticated parts by those small and medium-sized, technologically very sophisticated factories in Tokyo.

This linkage of parent factories and subcontractors is multilayered. It is common for an engineering firm to commit production of parts, even of the same parts, to several producers to raise the quality of parts and lower the cost by the competition among subcontractors. Parts producers also contract with as many makers as possible to stabilize their business. So, particularly in the Keihin machinery industry, characterized by its production of high-class products in small quantities and of varied kinds, many parts producers maintain their business by shipping small amounts of parts to many makers. Subcontractors (and makers) function in a complex linkage within groups of machine-makers who produce completed products with groups of parts producers.

This complex linkage may be classified into several types. The first is the linkage between machine producers and their parts producers, that is, the linkage within the same kind of industry. The second is the diagonal linkage with another kind of machine producer because of the technical features of the parts. The third is the subcontracted work of completed products. This is a type of production wherein large finished product suppliers commit the production of part of their output to small-sized finished product producers with their own brands. This method is common in the production of machine tools, electrical machinery, cameras, and meters. The large producers maintain their business by utilizing the finished products of subcontractors concentrated in the southern district. The fourth is the utilization of secondary subcontractors, and the parts producers organize their production through their own subcontractors. Given these circumstances, the southern district is still the center of machine production in Keihin region. The machinery producers who have large factories in the outer Keihin ring depend upon parts suppliers located in the southern district, which, therefore, is not only the center of parts production but also functions as the technical center for the machinery industry.[59]

The question must be whether these relationships constitute an innovative milieu. Toshiba's Yanagicho Works in Kawasaki suggests that they do. Known as a "brain factory," it manufactures an extraordinary range of different models of products with the aid of 164 supporting enterprises who supply some 65 percent of the total value of output, most of which are small and many of which are family firms. Most significant among these are the 92 *gaichu-san*, key factories that provide special order goods to factory specifications. Fruin shows how Toshiba's management works closely with these contractors to improve quality, constantly evaluating their performance on a complex scaling system. The relationship is described as one of co-partnership and co-prosperity, which in Japanese are synonymous.

Internalizing R&D: the Japanese model of innovation

Besides subcontracting, there is another distinct feature of Japanese industrial organization: long-term employment and hierarchical structures in the large corporations that constitute the primary labor market. This is more recent than many observers suppose: originally, the Japanese labor market was fluid and did not have any distinct structure. But, to overcome a lack of qualified labor, large corporations started to create internal labor markets in the 1920s, adopting a series of innovations including in-house training and promotion, firm-specific job classifications, and seniority-based payment. These features were refined and came to be widely practiced in the late 1940s and 1950s, through management's efforts to reassert control over wage determination and work rules against a militant labor movement immediately after World War Two.

The important point is their effect on patterns of Japanese R&D, which – in sharp contrast to the pattern of Silicon Valley – is heavily concentrated within the large corporations. Long-term guaranteed employment encourages these corporations to transfer researchers and engineers between their research and manufacturing divisions. This tends to increase the horizontal communication between laboratory and shopfloor, as well as between different departments. The scheme of remuneration, based partly on seniority, allows researchers and engineers to reap returns on human investment only when and if they are promoted to higher ranks – making it hard for personnel in midcareer, who possess a large amount of firm-specific knowledge, to quit that firm.

The Keihin region is the specific place where this distinctive industrial organization has brought about the unique evolution of the Japanese electrical, and then electronics, industry. The initial location of the industry in the region benefited from the external conditions of the region's innovative milieu during the Meiji period, which were mentioned earlier. Many small entrepreneurial firms were created in Tokyo, laying the foundations of the dense networks of small and medium-sized firms in the present day. As historical conditions gave rise to the distinctive institutions of Japan's industrial organization, an important part of that innovative milieu came to be internalized within large corporations. Those large corporations gained the flexibility to enter new fields by intra-organizational transfer of innovative resources. The movement of electrical engineering-based corporations into electronics was just such a case. At the same time, these large corporations rely on networks of small and medium-sized firms in the same region as a source of flexibility. Through their contracts with large corporations, small and medium-sized firms have raised their technological levels. This in turn has cumulatively tightened their linkages to large corporations. The evolution of the Japanese electric machinery industry – from an electrical engineering-based one to an electronics-based one – is a result of these interacting forces. The evolution is not only historically-specific but also spatially-specific to the Keihin region. It is perfectly logical that electronics-based industries should have emerged so strongly here.

There is little dispute that this distinctive industrial organization is well suited to downstream R&D activities, such as development research and engineering. Japan's strength in these activities, which took shape during the postwar period, precisely coincided with the rise and growth of its electronics industry. As electronics – in particular, semiconductor devices – became technologically mature and the basic technological framework became well-defined during the postwar period, downstream R&D increased its importance. But this very association could well be a disadvantage in the process of shaping the technological breakthroughs that will shape the next long wave.

Further, it may be a very specific product of Japanese society. Makoto Kikuchi, a leading researcher in semiconductor physics and solid state electronics, suggests that there are two modes of creativity: one is well suited to creating breakthroughs and setting a new technological framework, while the other flourishes within an already-established technological framework. Kikuchi argues that, through their social education, Japanese researchers are endued more with the latter mode of creativity than with the former. They exercised the latter mode, which Kikuchi calls "adaptive creativity," when electronics became technologically mature and its technological framework was already set by American researchers. If Kikuchi is correct, it is an open question whether the Keihin region is likely to continue to be a center of future technological breakthroughs.

THE NEW METROPOLES

MUNICH

The old-established global cities have shown extraordinary resilience in maintaining their position as high-technology innovative milieux. But, during the half-century since 1945, there have been two remarkable cases where such a metropolis found itself effectively supplanted by an upstart city far away. Both these cases, it must be said at the outset, are special. One, the displacement of Berlin by Munich, arose in the trauma of the end of World War Two, and seems at first simple to understand – though, examined in detail, it in fact poses unexpectedly difficult questions. The other, the superseding of New York City by Southern California, is a far harder phenomenon to explain – and, finally, here too the key proves to lie in the extraordinary pressures brought about by global conflict: first in World War Two, and second in the Cold War of the 1950s.

For the Germans, the Munich region is "Municon Valley": a journalistic invention of the 1980s, but one that expresses the city's dominant position in the German – and the European – electronics industry. And this in turn has made Munich, long viewed by Germans alternatively either as a place of artistic eccentricity and genteel living or as a city of beer cellars and storm troopers, into one of the country's leading industrial cities. Indeed, the first industrial city: the Munich region shares with West Berlin the title of the greatest industrial center of the former Federal Republic, and in recent years has overtaken both West Berlin and Hamburg in terms of numbers of industrial undertakings. Munich's preeminent position, so recently acquired, essentially springs from the city's "electronification" (*Elektroniserung*): its leading role as a center for manufacturing and trade in microelectronics on the German and indeed the European market.[60]

And this has been underlined in the years since the economic crisis of the mid-1970s, when the basic industries on which Germany based its postwar economic miracle – electrical equipment, vehicles, and chemicals – have ceased to drive the economy; the result has been a deepening north–south divide in the country's economy, separating the regions of older and heavier industry – the Ruhr, the Saarland – from the concentrations of the newer high-technology industries, above all electronics. And among these concentrations, the Stuttgart–Munich Corridor ranks overwhelmingly the first, with the biggest concentration of all (Figure 7.5) within 20 miles (30 kilometers) of Munich.[61]

Within the Federal state (*Land*) of Bavaria, at the end of the 1980s, the electrotechnical industry – as the Germans call it – ranks first with 245,000 employees, 25.5 percent of the workforce. Every fourth job in this industry in the pre-1990 Federal Republic of Germany was in Bavaria, and the state's share of total exports was 23 percent, and of output, 30 percent. Half of the German workforce of Siemens, one of the largest electrical firms in the world and Germany's largest private employer, was found in Bavaria in the early 1980s, 50,000 of them in Munich alone.[62] And it is the high-value R&D, planning and organizational activities that are especially concentrated here, though Munich is also an important center of electrical assembly, which is a major employer of the city's female workforce.[63]

Within the industry, electronics forms a special "critical mass": leading

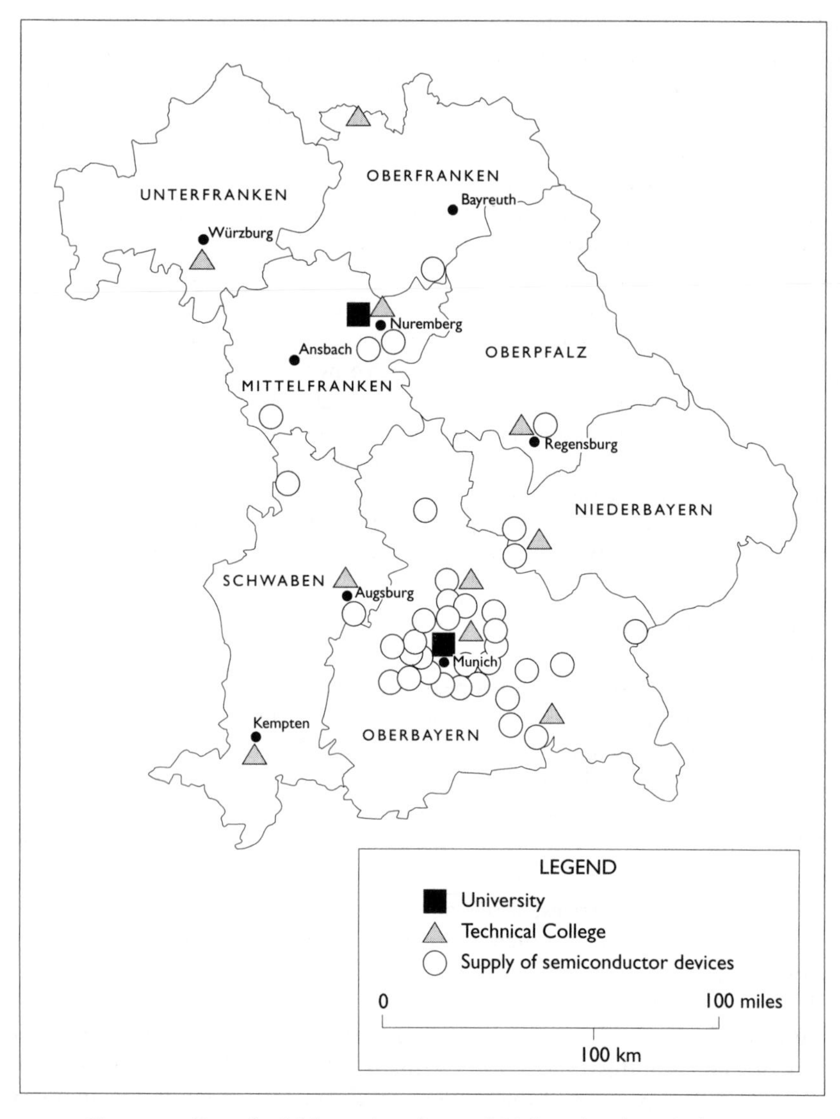

Figure 7.5 Bavaria: higher education and high-technology industry
Source: Bavaria, 1985

companies – Siemens, AEG, SEL, Diehl, Bosch and Telefunken – have located in Bavaria, above all around Munich. About 600 companies make electronics components and systems, and about 2,400 use them. Electronics firms have grown in number from 287 in 1975 to 3,008 in 1985, creating 28,000 new jobs. Of the new firms established in electronics since 1975, nearly half are in software, another quarter are making electronic systems and facilities for industrial use. Like the great majority of firms in Bavaria, they tend to be small and medium-sized.[64] One in every three German workers in data-processing, computers and related areas is found in Bavaria; about 40 percent of all German software firms are here.

This is perhaps the leading center of semiconductor production in Europe. Nearly all major semiconductor producers have their European headquarters or a major marketing outlet in Bavaria; major companies include Texas Instruments at Freising near Munich, Motorola in Munich, Siemens in Munich and Regensburg, Hitachi in Landshut and Landsberg, Fairchild in Wasserburg, Semikron in Nürnberg, and Eurosil in Eching near Munich, and the industry employs around 10,000 people. The world's largest silicon producer, Wacker-Chemitronic, is headquartered at Burghausen in Bavaria. There is a large concentration of small and medium-sized subcontractors. The major users of microelectronics – in mechanical engineering, vehicles, office, data-processing and communications technologies, aircraft and aerospace, precision and optical engineering – are here or in neighboring Baden-Württemberg.[65]

Particularly in recent years, this area has also become a major producer of software. According to a 1985 survey, there were some 650 software houses in Bavaria with sales of DM1,000 million: three-quarters were in standard applications software (24 percent), individual applications (27 percent) and dedicated applications (23 percent). Two-thirds concentrate on just one main activity. They reckoned with sales growth of 20 percent per annum through the remainder of the 1980s. Only 31,000 out of 334,000 Bavarian enterprises used data processing in 1985, the same proportion as in the Federal Republic as a whole. The software houses are unevenly distributed, with 54 percent in and around Munich, and they tend to continue to cluster there because they are spin-offs.[66]

Because of the concentration of demand, there has been a huge growth in the number of qualified graduates in the field of electrotechnics, with growth rates 30–100 percent above the federal average, and an even bigger jump in the output from technical colleges: state-licensed electrician graduates jumped by 170 percent in two years.[67]

In Bavaria, some 75 percent of high-tech industry is concentrated in Upper Bavaria, of which, in turn, no less than 75 percent is located in and around Munich. Within this area it is fairly evenly distributed, though perhaps with a concentration near the airport: local concentrations are found in Kaufbeuren (telecommunications equipment), Landshut and Landsberg (Hitachi), Nördlingen (Fischer), Neuperlach and Augsburg (Siemens). A large number of firms are not included in the figures on which this map has been based; they go under different names. These are often the biggest, best-known firms. The analysis also does not show high tech embedded in different industries like automobiles, although a great deal takes this form. Within Munich itself, the electrical engineering industry accounts for just

Table 7.6 Munich: manufacturing industry, 1985

Major sector	*Employment average*	*Turnover (DM 000s)*
Basic materials	9,689	5,109,467
Producer goods	134,619	40,542,471
(Electrical)	(51,228)	(16,945,308)
Consumer goods	16,484	3,026,691
Food & drink	11,493	4,242,228
Total	172,285	52,920,862

Source: Klingbeil, 1987a, based on *Statistisches Jahrbuch der Stadt München 1985/6*
Note: Totals may not agree due to rounding

under 30 percent of all manufacturing employment, and just over 30 percent of all manufacturing turnover (Table 7.6).[68]

Surveys show that no less than 2,000 electronics manufacturers and trading firms are found in the region, which extends out to include major cities like Augsburg, Landshut, Rosenheim, Wasserburg, Burghausen, and Regensburg. They include leading American and Japanese firms such as Digital, Hitachi, Sanyo, Murata, Motorola, Texas Instruments and Fairchild, as well as major German producers like Siemens and Nixdorf. Hardly any major semiconductor producer can afford not to establish either its European headquarters or at least a major branch activity in Munich or Bavaria. Round these major producers has developed a dense network of small and medium-sized firms which act as subcontractors and play an important role in development and applications.[69]

Many research institutes, both inside and outside the city's universities – such as the Max Planck Institute for Plasma Physics in Garching, the *Gesellschaft für Strahlen und Umweltforschung* (Society for Radiation and Environmental Research) in Neuherberg and the *Versuchsanstalt für Luft- um Raumfahrt* (Institution for Air and Space Travel) in Oberpfaffenhofen, all three on the outskirts of Munich – and no less than 13 institutes and departments of the *Frauenhofer Gesellschaft zur Förderung der angewandeten Forschung* (Frauenhof Society for the Promotion of Applied Research), testify to the close contacts between research and industry. College graduates more readily find well-qualified work opportunities than in other regions. The region has the strongest attraction to venture capital in all Germany, partly because of the wealth of regular trade fairs which offer an overview of supply and demand developments on the world market. The spatial concentration of large and small firms, R&D, fairs, and exhibitions provides a general infrastructural basis for the "electronification" of the economy, in a way not observed anywhere else in the pre-1990 Federal Republic, and not likely to be in the foreseeable future. On the demand side, the industry here is dominated by branches with important applications: office and computer technology, support systems, and military electronics.[70]

High tech moves out

However, the city's share of electronics firms and jobs is falling, as firms relocate to the suburbs, and others newly set up shop there. And this is only a microcosm of what has happened in the economy generally. Munich still dominates the economy of its region, because of the lack of competing centers: in 1971 it had 77 percent of total regional secondary- and tertiary-sector employment, but this share fell during the 1970s as jobs suburbanized.

This has been particularly marked for the high-technology sector. Even from the mid-1950s onward, with the development of the Cold War economy, the suburbanizing trend received a boost from major air and space firms like MBB, IBAG, Dornier and DFLVR. Multinational firms, like Texas Instruments, choose freestanding suburban locations just as they do in the United States. But local firms began to move too in the 1970s, and from that time not just production, but also a higher proportion of R&D, sales, and administration, began to move out. Needless to say, the process was very differently regarded by Munich and by the suburban communities.[71] But there was also a great deal of relocation within the city itself, most of it quite short-distance.[72]

The origin of the "Munich phenomenon"

Munich's extraordinary rise to becoming high-tech capital of Germany, and almost certainly of Europe, has little to do with any conscious planning. Least of all had it anything to do with deliberate regional policy. Officially, the German Federal Government has never encouraged the growth of high-technology industry in the region; on the contrary, its policy has been to support industrial development in remote border zones and, more recently, in the depressed older industrial areas. Munich's climb to greatness, we were told in an interview by local officials, has been a matter of free-market forces. More closely, it has been brought about by a rather extraordinary conjunction of global historical-political forces, economic forces, and governmental policies that have sometimes worked in contradiction to each other.

Munich was a relative newcomer among German cities: it was founded in 1158 by Henry the Lion near to the site of an older market established by the Bishop of Freising. From 1255 to the collapse of the monarchy in 1918, its fortunes were inextricably tied up with those of the ruling house of Wittelsbach. Throughout the Middle Ages, it was bypassed by the major trade routes, which ran east–west along the river Danube to its north, north–south through Augsburg to its west. The bourgeois oligarchy was conservative and complacent; the Counter-Reformation, which ran strongly in this Catholic part of Germany, was hostile to the development of science.[73]

But, soon after 1800, the city's fortunes changed. Under the Napoleonic "New Order," Bavaria was enlarged to become the third state of central Europe after Prussia and Austria; as its capital, Munich now competed directly with Berlin and Vienna. King Ludwig I's ambition was to make it a truly European capital, which he did by centralizing here the bishopric, university, and art collections; with the explicit aim of making this a major European artistic center, he attracted artists to the city. His son, Maximilian II, did the same for science professors; in turn, his successor, Ludwig II, attracted musicians. Thus Munich developed as a many-sided cultural city, especially in the inner suburb of Schwabing, which became a unique haven of avant-garde ideas: Germany's Montmartre.[74]

But Munich developed into something more. After 1840, because it was the capital and most important center of population, the city became the leading railway center of southern Germany, serving international trans-Alpine traffic.[75] Thus its long isolation from trade routes was ended. Now, at long last, it could industrialize. But, in the "Foundation Age" (*Gründerzeit*), the years of rapid industrial growth following unification of Germany in 1871, the city's lack of raw materials and water access caused it to concentrate on lighter, higher-value-added production. The wealth of education and research capacity led naturally to precision instrument manufacture of many kinds: machinery, machine tools, precision measuring apparatus, and photographic and optical goods.[76] In particular, during this period, the union of research and crafts traditions generated a very strong tradition in fine optics, represented by firms that became world-famous – Arnold and Richter, Linhoff, Rodenstock, Steilheil, and others – out of which, in turn, developed special skills in the manufacture of precision measuring instruments.[77]

Yet, even as late as 1925, Bavaria was still not truly industrialized: among 28 *Länder* and provinces of the German *Reich*, Bavaria stood at nineteenth place in terms of industrial establishments, a position that represented no improvement on the previous industrial census of 1907.[78] And most Munich industry was small-scale: in 1907 the city had only six undertakings with more than 1,000 employees, by 1925, 23.[79] Only 37 percent of the city's workers in 1907, and 38 percent in 1925, were in all forms of industry including crafts and homework.[80]

Nevertheless, in the course of the nineteenth and early twentieth centuries, fortified by the development of industry and trade, Munich grew rapidly. Its population exploded, from 90,000 in 1840 to 170,000 in 1871, and to 500,000 in 1900; thence, it grew a further 70 percent, to reach some 840,000 at the outbreak of World War Two.[81] After World War One the city gained from Vienna's decline, but could not compete with Berlin either as artistic milieu or as industrial center. It developed motor and aircraft industries, which gained from rearmament in the Nazi period.[82] But, down to World War Two, it was still not very significant as an industrial city in comparison with Berlin, which between 1850 and 1914 – even 1939 – was "Elektropolis": the greatest high-tech industrial center in the world, the Silicon Valley of its day.

Thus, Munich was a direct beneficiary of the collapse of the Berlin electrotechnical industry at the end of World War Two. For at that point, with extraordinary speed and suddenness, the industry left the former capital and relocated in the southern part of the new Federal Republic, above all in and around Munich.[83] By 1967 Berlin accounted for only 10 percent of electrical engineering employment in the Federal Republic against 24 percent in Bavaria or 20 percent in Baden-Württemberg, the two areas that had benefited most from its decline. Within Munich the electrotechnical industry employed 6.9 percent of total industrial employment in 1950 but 14.2 percent in 1968, an increase of 105 percent.[84]

In retrospect, it is not surprising that Berlin should lose its leading role; what is harder to explain is why Berlin's loss should be Munich's gain. Berlin in the nineteenth century had provided the first example in world history of a modern military-industrial state: an industrial region where state military interests worked in close conjunction with large private companies to generate industrial innovation. In the period 1945–9, these ties were severed as a result of Berlin's physical and industrial collapse, the subsequent division of the city, the establishment of the Federal Republic, and the effective end of Berlin's capital city functions. During the war itself, the Siemens company had been compelled to decentralize production out of Berlin to a host of factories all over the country; from winter 1943–4, the Siemens works in Berlin were hit by increasingly heavy air attacks.[85]

Just before war's end, in February 1945, the directors of Siemens had effectively decentralized operations to a number of regional headquarters in southern Germany, with Siemens & Halske in Munich, under Ernst von Siemens, at the apex: "Director Ernst von Siemens is appointed to guard all the interests of the house including the daughter companies in trusteeship" was the official formulation of February 19, 1945.[86] The firm's Berlin headquarters was effectively transferred to Munich, at first in various rented premises in the city center, eventually to a large works complex in the suburb of Obersendling with a palatial central headquarters in a classical house in the Wittselsbacher Platz, acquired on a "provisional" basis in 1948, and effectively made permanent in late 1949 after it had become evident that the country was divided.[87] Immediately after capitulation, Berlin's industrial capacity was effectively dismantled by the occupying powers; Siemens and AEG effectively lost close on 100 percent of their capacity.[88] The magic circle of contact between industry and scientific research – despite the continued prestige of Berlin's Technical University – was broken.

Why should Munich have been the beneficiary? The reasons are still shrouded in some degree of mystery. Carl Friedrich von Siemens, who headed the firm from 1919 until his death in 1941, had received part of his education at the Technical High School in Munich, where his mother had gone to live after his father Wilhelm's death; he retained a lifelong love of Bavaria, and had a mountain retreat in the Längau area near Ruhpolding, some 140 kilometers (90 miles) southeast of Munich; he began to decentralize part of the firm's operations out of Berlin in the 1930s;[89] Siemens had been represented there before World War Two.[90] And Munich did have attractions for this kind of industry: its lack of an older heavy-industrial legacy, with all that implied in terms of environmental quality; its tradition of crafts-based precision engineering; the excellence of its university education; its nearness to Alpine recreation; and its many cultural amenities and splendid public buildings.[91] These were all assets, but were shared in only slightly lesser measure by other southern German cities. Later, when microelectronics developed, it was perhaps fortunate that 50 percent of the raw material, silicon, came from Burghausen in Bavaria; but, because of its extreme lightness and value, this could not possibly represent a real locational pull.[92]

However, Munich benefited from some rather unusual circumstances. According to one unauthorized history of the Siemens firm, at the end of 1944 a Siemens agent in Sweden – a neutral country in World War Two – obtained a secret copy of the map resulting from the Yalta agreement between the great powers, showing the planned division of Germany. This map, smuggled into the firm's Berlin headquarters, indicated clearly to the Siemens management that they had no future in the Russian zone of occupation. It was at that point that Ernst von Siemens, Carl's son, was ordered to Munich to set up an organization there, and the firm was put under his trusteeship – a move that proved propitious, since the then head, Carl's nephew, Hermann, was interned on suspicion of war crimes in 1945 (though soon released).[93] Immediately after the end of hostilities Bavaria's Nazi wartime command economy structure was allowed to continue in operation, rather remarkably, by the American military occupation; and this set out systematically to industrialize the province.[94] A stream of refugees, mostly well-qualified – from the textile and machine building industry of Saxony, the printing industry of Thüringen, and the export-oriented industries of the Sudetenland – brought the reestablishment of many firms here, and so quickly built up a high-quality industrial workforce. The high political risk in West Berlin led to relocation, not only of the electric industry, but of other high-technology lines. Labor mobility, in the growth years of the 1950s and 1960s, led to a north–south migration, at the very time that the opening up of Europe was widening the market. Munich's population in 1950 was 843,000, slightly less than in 1939; by 1970 it had shot up to 1,312,000, after which it more or less stabilized.[95] And all this in turn led to agglomeration effects, with a range of small and medium-sized ancillary and service industries; the high-technology armaments industry concentrated here in the 1960s.[96]

A regional policy? Bavaria's education and research support

There was however one important respect in which government policy played a role; and, just as in the case of London, it was an anomalous and even perverse one. The subventions of the Federal Ministry for Research and Technology to the region were especially high.[97] One analysis concluded specifically that the aims of German regional policy had played a minor role in R&D funding; there had been only the most limited coordination between regional aims and those of other policies.[98] Under the new federal research policy to support R&D in small and medium-sized firms, which came into force in 1978, there is direct payroll support plus support for external contract research, encouragement of pilot projects and testing, and innovation-oriented regional support policies; the main beneficiaries have been the newer-industrialized areas of the south (Table 7.7).[99]

Further, the State (*Land*) Government actively supports education and research. Ten percent of the *Land* budget goes on higher education and research. Bavaria, unlike other *Länder*, makes a distinction between fundamental (university) research and applied research, which is concentrated in Munich. It has nine universities receiving DM3.8 billion per year (1986); in addition, there are *Fachhochschulen* (Technical High Schools) and about 400 research institutes. The Technical University of Munich has seven chairs in microelectronics, the University of Erlangen-Nürnberg has 10.[100] No less than 63,000 students study maths, natural science, and engineering in Bavaria's universities and technical colleges, a growth of 30 percent in three years. There is a trend toward the formation of special subject universities which the government will be more likely to support.

Southern Germany, including Bavaria, has also benefited from federal policies regarding the location of major fundamental research institutes. These institutes are funded 90 percent by the Federal Government, 10 percent by the *Länder*. Fig. 7.5 shows an extraordinary concentration in the prosperous south, above all in Munich and in the Stuttgart area.[101] There are 35 institutes in the former West Germany, of which six are in Bavaria,

Table 7.7 Federal Republic of Germany: research support, per cent, 1985

Type of area *Program*	*Direct project support*	*Personnel subsidy*	*Contract research*	*Compare: population*
Old industrial	14.3	5.7	5.1	10.9
New industrial	71.5	50.3	54.1	44.6
Urbanizing	8.7	29.7	25.5	28.7
Rural	5.5	14.3	15.3	15.8
Total	100.0	100.0	100.0	100.0

Source: Grotz, 1989: 268

12 in Baden-Württemberg, two in Berlin, six in Nordrhein-Westfalen, one in Bremen, one in Saarland, none in Hamburg, two each in Lower Saxony and Hessen, and none in Rheinland-Pfalz. All the 11 prestigious Max Planck Institutes are located around Munich. Other important research centers in Germany are Freiburg, Tübingen, Köln-Düsseldorf and Berlin; Göttingen, Hannover and Hamburg are not important, and the last has an image problem; the Ruhr, it is felt, will become attractive in another 10–15 years. Ordinary professors do not leave because they find Bavaria too attractive.

Bavaria has not left matters to chance. Because state policy now stresses microelectronic research, it has departed from the usual 90:10 formula by paying 100 percent of the cost of a new Frauenhof Institute in Erlangen near Nürnberg, which came into existence in 1990 at a total cost to the Bavarian Government of DM100 million. Since the existing Frauenhof Institute for Solid State Technology, with a staff of some 65, is sited in Munich, this clearly reflects an attempt on the part of the *Land* government to spread microelectronic R&D into other regions.[102]

These institutes have a great deal of autonomy; they have close connections with the universities which are often next door, and their directors tend to have university status. The *Land* government leaves policy direction to the research director, as in a university. Most institutes, we were told by Technology Ministry officials, are successful both in terms of research and industrial application. But the Ministry would clearly like to bring universities, institutes and industries closer together; it feels that not enough research flows to industry. In our interviews, we asked: is there much spin-off? There was some, we were told: younger workers, who get their Ph.D.s in the institutes, may leave after 3–5 years there, to go into industry. But there is a reverse movement from industry into research, too.

But there was no other specific encouragement. Quite unlike most other places considered in this book – though again, quite like London – high-tech growth in and around Munich has not come about through the deliberate establishment of science or technology parks. There are only three such in Bavaria – in Munich, Nürnberg and Würzburg – and even these are not really science parks in the sense usually understood: the real high-tech industries are not developing there.

SOUTHERN CALIFORNIA

Southern California is a land of cultural innovation and entrepreneurial dynamism, with a history of rejecting the past and realizing its own dreams of the future. It is also a place of striking paradoxes. Not the least of these paradoxes is the fact that for half a century since 1940, this place, originally a nest of free-enterprise spirit, has based its remarkable economic growth on a defense-driven high-technology manufacturing complex, entirely dependent on the support of the Federal Government. The "Warfare State," not Hollywood or Disneyland, is the foundation on which the Southern California economy has built its huge prosperity. On this base, despite the fact that it directly employs less than 8 percent of the Greater Los Angeles labor force, entire layers of markets and supplying industries have spun off. An extraordinary combination – bold entrepreneurialism in the 1920s and 1930s, followed by explosive high-technology defense demand, and dramatic technological innovation in the aerospace and electronics industries – has transformed Southern California into one of the two largest territorial complexes of high-technology manufacturing in the world (the other being Greater Tokyo), and certainly into the most important military industrial complex of modern times.[103]

If we accept for working purposes Allen Scott's functional definition of the Southern California megalopolis,[104] it stretches for about 250 miles (400 kilometers) along the coast, from Santa Barbara to San Diego, with its central nucleus in Los Angeles and its most rapidly expanding zone in Orange County. This megalopolis includes 145 municipalities in seven counties, from north to south: Santa Barbara, Ventura, Los Angeles, San Bernardino, Riverside, Orange County, and San Diego. In 1988, the area had a total population of 16.3 million and a total *manufacturing* labor force of 1.4 million. In 1986 it contained 2,173 high-technology firms employing 435,500 workers, a figure that had increased by almost 50 percent in 10 years.[105] A narrower geographical definition would in fact increase the density of this concentration: Los Angeles County alone employed about 260,000 high-technology workers in 1986, and Orange County about 90,000. The manufacturing activity in which this complex originated was primarily the aircraft industry, to which was subsequently added the natural expansion into guided missiles and space vehicles. Communications equipment, electronic components, and computers arrived in the 1960s and 1970s to complete the picture, as aerospace firms both internalized the emerging technologies, and also created markets for new firms in defense-related electronics, specialty products, and technological services.

This agglomeration does not however constitute a single industrial district. Allen Scott's work has established that there is in fact a constellation of several technopoles, constituted in a historical sequence over the last 60 years or so.[106] Three predate War World Two in their original development: the west-central area of Los Angeles around the site of Los Angeles International Airport, representing the original hub of the aerospace industry, and stretching from Santa Monica and Culver City in the north to Inglewood, El Segundo, Hawthorne, and Torrance in the south; the Burbank-Glendale area; and the San Diego Bay area. Then, starting in the late 1950s, and still in process in the 1980s, a movement of manufacturing decentralization created new defense-related technopoles; the most important of which, by far, is Orange County, particularly around Irvine and its surrounding area. Other important developments include the Chatsworth-Canoga Park and Van Nuys areas in the San Fernando Valley, Kearney-Mesa and La Jolla–Sorrento Valley, north of San Diego, and Palmdale in the northern fringes of Greater Los Angeles, around the Edwards Air Force Base. Although each one of these technopoles forms a relatively autonomous complex of manufacturing and services, there are subcontracting and market relations throughout the whole region, making it a true network of technopoles, often connected through divisions of the same large firm and via the shared networks of small and medium-sized companies which serve several of these giant aerospace and electronics firms (Figure 7.6).

This extraordinary regional production system was the offspring of a most unlikely concubinage between bold industrial entrepreneurs, innovative scientists and engineers, and farsighted Pentagon warriors, who together, in the 1930s, forged the Los Angeles aircraft industry. This infant industry exploded as a manufacturing powerhouse during War World Two, contracted immediately afterwards, but then massively expanded in the 1950s and 1960s as it effectively created the new aerospace and avionics industries whose success would ultimately prove decisive in the Cold War. This extraordinary technological and industrial saga actually goes back to the first major air show in America, held in Los Angeles in 1910, in the area that was going to become the heart of the "Aerospace Alley," north of Long Beach. Yet, until the 1930s, the bulk of American aircraft production was concentrated in the Northeast and Midwest. The gradual shift to Los Angeles, right down to the moment when the Western metropolises became the hub of the industry just before World War Two, was due – as usual – to a combination of locational factors, entrepreneurial initiatives, and synergistic relationships that take place in the formative years of any innovative milieu.

Los Angeles did have some attractive locational factors: mild weather year-round, enabling aircraft tests and, more importantly, outdoor manufacturing work on large metal structures; large tracts of undeveloped, cheap land in the periphery of the urban area; and, very early, an excellent educational and research program in aeronautical engineering at the California Institute of Technology, started by physicist Robert Millikan in 1925 with the help of the Guggenheim Foundation. Four years later, Millikan recruited one of the leading German scientists in aeronautics, Theodore von Karman, to head a new Guggenheim Aeronautical Laboratory that would spawn, in 1943, Cal Tech's Propulsion Laboratory, the world's most advanced center in aviation and space research.

Yet, while these elements would become very significant as parts of the complex that would ultimately emerge, the origins of the Los Angeles aerospace industry can be traced back directly to the early location of a few start-up companies, supported by civic boosterism on the part of the Los

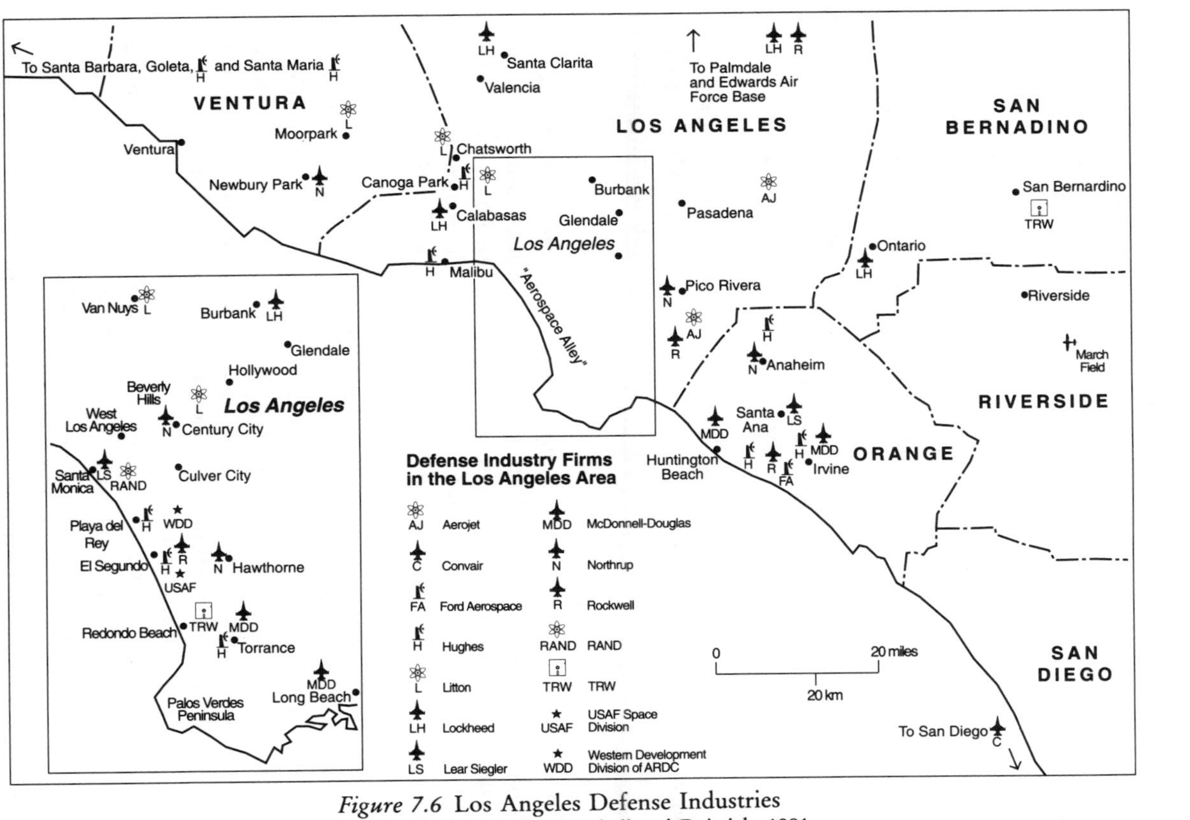

Figure 7.6 Los Angeles Defense Industries
Source: Markusen, Hall, Campbell and Deitrick, 1991

Angeles business community, and by a few innovative military commanders. In 1909, Glenn Martin (the ancestor of Martin-Marietta) built his first plane in Santa Ana, Orange County. In 1920, Donald Douglas, the co-founder of McDonnell Douglas, set up an aircraft company in Santa Monica. He subsequently recruited a brilliant aviation designer, Jack Northrop, who headed a separate division in El Segundo in the 1930s before founding his own company, a future mainstay of the defense aerospace industry. Two brothers, Allan and Malcolm Loughead, created the Lockheed Company in a Hollywood garage in 1926. In 1935, North American Aviation (later to become Rockwell) moved its base from Baltimore to Inglewood, next to the Los Angeles Airport. And in 1934, Howard Hughes, who had some success at making movies, was so impressed by the aviation drama he had produced four years before[107] that he hired a Cal Tech graduate to start a sports plane company in a leased corner of a hangar in Glendale. Much later, after World War Two, he hired two gifted engineers, Ramo and Wooldridge, who escaped from his erratic management in 1953 to found TRW, one of the most innovative high-tech defense companies.

Thus, the line of development is clear, even in terms of individual names, from the entrepreneurial start-ups of the 1920s and 1930s to today's aerospace giants. But why should such a concentration of entrepreneurial and engineering talent in aviation occur precisely here in Los Angeles, and not in some other place?

There seem to be several reasons. First of all, the pioneers received considerable help and support from Los Angeles business and civic groups. The most prominent figure was the publisher of the *Los Angeles Times*, Harry Chandler, who – convinced of the future of the industry and of the critical role Los Angeles could play in it – personally supported and funded several aviation start-up companies. Together with a group of citizens, led by real-estate developer Harry Culver (the founder of Culver City), in 1930 they campaigned to persuade the city of Los Angeles to build a Municipal Airport, eventually to become today's Los Angeles International Airport, as a basic facility for the development of the new industry. Indeed, the Los Angeles City Council pursued an active policy of economic intervention in key areas of infrastructural development, that proved critical for the city's future as a major manufacturing center.

Second, in addition to the civic groups, the Los Angeles-based military establishment also played an important role in the early stages of the aviation industry. Another key figure in the unfolding saga was "Hap" Arnold, the commander at March Field near San Bernardino during 1931–6, who learned about the technological developments in aviation from his friend Donald Douglas. In 1938, through Arnold's support, the Army commissioned the Cal Tech Jet Propulsion Laboratory to do research on long-range jet-propelled missiles, and eventually built for it $3 million worth of new facilities. Arnold later became Chief of the United States Air Force. In 1944, he asked von Karman and the Cal Tech team to explore the future of military aviation. The committee's secret report, written in 1945, envisaged long-range ICBMs and earth-orbiting satellites, and became the foundation for the strategic program of the Air Force over the next decades. Help for this work came from a civilian research center established in Santa Monica by Donald Douglas, with the support of the Air Force: founded in 1946, its name was RAND, for Research and Development.

However, all this support from different quarters would have had no result without the presence of the engineers and entrepreneurs who showed up precisely in Southern California to indulge their dreams. What was it about the place that attracted them so deeply and so consistently? It may be, just may be, that there is a third critical dimension to the Southern California experience, namely its cultural openness. An open land, an open society, on the western edge of the Western world, even more so than its more traditional northern rival in the San Francisco Bay Area, Los Angeles has always been a "city of dreams," Borges's "Aleph", where all experiences come together, in the image suggested by Edward Soja.[108] As Markusen, *et al.* write,

> it is surely no accident that two of the new technologies that have captured America's imagination in the twentieth century, flying and cinema, were brought to maturation in Los Angeles . . . Los Angeles was a place where people came to leave behind their old ideas about cities, industries, and culture. The unique culture of the early twentieth century gave Los Angeles a special momentum, which it has exploited to this day.[109]

After the foundations of innovation were thus established, the wars – hot and cold – did the rest. During War World Two, Los Angeles became the powerhouse of the American aircraft industry, and this complex would ultimately decide the fate of the war: in 1939 the aircraft industry in Los Angeles employed 15,000 workers, in 1943 no less than 190,000. Then, in the late 1950s, the United States escalated the military technological race with the Soviet Union; and, as the aviation industry naturally expanded into missiles, satellites, communications, and avionics, it did so from a Southern Californian base. Three main factors seem to account for the overwhelming preference for Southern Californian companies on the part of the defense establishment: the technological excellence of these companies and their supporting research network, both private and public, in the industry and in the university; the intricate web of commercial and technological links in the area, together with the concentration of a gigantic pool of highly-skilled labor specialized in the trade; and the long-time personal and corporate ties between Los Angeles-based companies and the top officers of the United States Air Force, the branch of the armed forces that won the critical competition between the different services to lead the missile and aerospace program.[110]

Boosted by an extraordinary level of demand from the military quarters, the original technopole located around the Los Angeles Airport expanded, branched out, and began to subcontract. As Scott has shown, these firms practised vertical disintegration and horizontal networking in order to cut costs and develop a flexible response to variations in demand.[111] Thus, a growing number of small and medium-sized firms were set up to serve the giants of aerospace, as simultaneously the latter diversified at speed into electronics and communications, setting up new divisions with a great deal of autonomy. Each new contract brought a number of subcontracting arrangements that created opportunities for the birth of new firms. Once these were in place, they networked with others, in an endless process of cross-fertilization and market expansion. While major corporations tended to expand in areas controlled by their establishments, almost always inside the Los Angeles Basin, small and medium firms clustered in new, high-density industrial districts, taking advantage of agglomeration economies, and pushed into proximity by the need for frequent contacts to secure customized production and services. Thus, along the southern freeway corridor, Orange County developed, bringing a new wave of decentralization of the existing technopole with every new round of technological innovation and military buildup.

The synergy created in such an innovative milieu, along with the sheer size of the production complex, ensured the permanence of the aerospace and military electronics industry in Southern California. And this was in spite of the rapid deterioration of the quality of life, under the pressures of unplanned industrialization and urbanization processes, that ultimately transformed one of the world's most beautiful urban locations into a locale for a science fiction horror movie, as the "city of dreams" became the "suburbs of nightmares."

Thus, the defense technopoles of Southern California first emerged from entrepreneurialism and innovation linked to a new kind of urban culture; but, almost from the start, they were decisively supported by a state that went on to build the most formidable industrial arsenal of all times, a high-technology complex so powerful that it will survive its origin, even if the peace dividend ever becomes a tangible reality. In this interactive process between an innovative city, an innovative industry, and a supportive Warfare State, the Los Angeles-centered megalopolis emerged as a new spatial form – a form characterized by endless decentralization of the various technopoles, the productive building blocks of the system, and a resulting polycentric structure linked by the network of freeways. Through its media industries Southern California went on to project this image, massively and successfully, to the entire world; but behind it, at the core of its real estate and advanced services businesses, the defense industry provided the real dynamo of growth.

SUMMING UP

What is clear is the remarkable resilience of the old metropoles. They were the original innovative milieux, a position they won early in the industrial era through their preeminence in customized precision engineering. They parlayed this, in the critical decades of the 1880s and 1890s, into the new field of electrical engineering. At that time, buoyed by the market, a few firms, which had begun lives as small workshop undertakings, rapidly transformed themselves into major industrial corporations. As they did so, they migrated outward in search of land. They consolidated their power as producers of both capital and domestic consumer goods during the 1920s and 1930s. In both London and Tokyo, they became critical centers of the World War Two production effort. Then, during the 1950s and 1960s, they diversified into electronics, and further decentralized toward the peripheries of their respective metropolitan regions.

This was not entirely a spontaneous process. In all three cases, the state played a crucial role for reasons of defense or national prestige: in Tokyo during the 1880s and again in the 1930s; in both London and Paris, in the 1950s and 1960s. Particularly in both London and Paris there was a remarkable circumstance, which could hardly be coincidental: a cluster of defense-related and state-promoted research activities developed, in an area toward the southwest corner of each extended metropolitan area, on land that was historically very prestigious because of regal associations. And around these, by deliberate design, groups of specialized high-technology research or production developed. Indeed, the same southwestern bias is evident in Kanagawa prefecture south of Tokyo, though here there is no corresponding state effort; there, as seen in Chapter 4, the governmental research institutes were relocated north of the capital.

These efforts seem to have been sometimes well-conceived, sometimes not. In particular, in the period since World War Two, neither London nor Paris seems to have been successful in maintaining its position as a genuine innovative milieu in the sense of demonstrating true synergy between the parts. London has been locked into a specialized system of defense contracting which seems to have inhibited its participants in going out to the marketplace; Paris has encouraged its high-tech private sector to seek large state contracts with the aim of enhancing the prestige of France. But in both, government laboratories and private contractors remain largely separate, even though one may contract to the other; there is only a limited degree of spin-off or synergy.

Tokyo is the outstanding exception, and is by any token one of the outstanding innovative regions of the twentieth century. Yet the nature of its achievement remains elusive. It has been attained through a distinct, indeed unique, combination of three elements: organized R&D within hierarchically-organized large corporations, systematically driving these companies toward the leading edge of applied research; competition among technically-sophisticated subcontractors; and government coordination through a long-term program of national technology development. The first stands in sharp contrast to the Silicon Valley model; the second – though not so far, apparently, much-researched – may offer greater similarities; the third does not suggest many parallels, save distantly in the form of defense contracting in the early days of Silicon Valley.

The conclusion that compels itself is that in the late twentieth century, in all three cities, we find a form of corporate or statist innovative milieu, save that in London it is relatively unsuccessful, in Paris only partially successful, but in Tokyo outstandingly so. The reasons for that, like so much else, seem to lie deep in the Japanese collective culture, in particular in the seamless web of public–private and private–private cooperative agreements that appear to pervade the entire economy.

Both the two upstart metropolitan cities represent some kind of mystery. No one in, say, 1935 could have predicted that either place would supplant the established center. Munich was the legatee of the sudden demise of Berlin, the greatest high-technology electrical production center on the European mainland; but why, precisely, is not easy to see. Perhaps, indeed, highly personal factors played a role: apparently, the senior figure in the Siemens concern, in 1945, had been educated in Munich and had particularly warm feelings towards the place; and, once Siemens was established here, perhaps the rest logically followed. It cannot have been other than helpful, either, that Munich was a major center in the American zone of occupation, and so may have benefited from a concern to move critical strategic industries away from the Russian sphere. There is, no doubt, a story here that has not been told. Whatever the case, the subsequent Munich story is a familiar one of circular and cumulative causation. Once the critical structuring decisions had been taken, the rest fell into place.

Nevertheless, the fact is that in this game of historical accident, Munich was well-placed as a contender. Like London, Paris, and other great European capitals, it had developed both a rich tradition of scientific education and research, and a parallel structure of highly-skilled craft industries. It thus could offer, to perhaps a greater degree than any other competing West German city, what is now called the soft infrastructure of accumulated human capital. So, in retrospect, the events followed a logical enough course.

They did so quite early in the history of the fledgling Federal Republic, at a period when events were in rapid flux and reflexive regional policy was not yet well-developed. So it is not at all surprising that, once such a policy was in force, it proved virtually impossible to overcome the built-in momentum that was driving the Munich economy forward. In any event, the truth may well have been that when it came to hard choices, regional policy took a bad second place to other considerations deemed more significant: national technological development, Germany's position within the European Community, national and international defense priorities.

So, with the benefit of hindsight, it is not difficult to follow the story. It seldom is. But in those chaotic months of 1945, it would have been difficult indeed to predict the eventual outcome. Historical accident may have played an important role. But there may have been some deeper agenda that had to do with global strategic advantage.

The same is more certainly true of Southern California. Until World War Two, and indeed for a few years after that, indeed much later, the metropolitan innovative milieu appeared to be concentrated where it had long been, in the corridor or axis that led from Boston to Philadelphia, but was particularly concentrated in and around New York City. This area – the crucible of the American electrical revolution of the 1870s and 1880s – still seemed, as late as 1950, to exercise a continuing dominance over the emerging field of electronics: it had an established radio industry, rapidly moving into television applications, and a long tradition of innovative research at Bell Laboratories in New Jersey. Even as late as 1945–9, the major fundamental advances were being made here: the transistor was invented at Bell in December 1947, while the first computers were being developed in nearby universities like Princeton and Pennsylvania.[112] If anyone had cared to predict the venue for the forthcoming marriage of space travel and electronics, surely it would have been somewhere in this corridor that joined New York, the Bell Laboratories at Murray Hill, New Jersey, and nearby Princeton University.

Yet somehow, the Northeast Corridor lost this established lead. As already seen in the case of Silicon Valley, some of this could be the result of chance factors. But in the case of Southern California, it seems clear that the process was more systematic than that. The city had aggressively wooed the founding fathers of the industry, who had then emerged almost from nothing to create a major industrial base during World War Two. It had also established a crucial research base at Cal Tech, with key military and industrial links. These proved crucial when the critical decisions had to be taken, in 1953–4, on the location of the core of the Cold War defense production complex. There was a chain of circumstance, which at each stage gave Southern California further built-in advantages. Nothing succeeded like success.

There is a strange similarity to the Munich story. There the sequence is clearer, because the original Berlin base of the industry was physically destroyed, and because it appeared too strategically dangerous to rebuild it in the heart of the Soviet occupation zone. Given those basic facts, the industry had to be rebuilt somewhere, and Munich appeared to be as good a location as any. But there, too, there was a hidden story in the Bavarian wartime economic command, which the American occupying forces seem to have permitted, and even encouraged, to remain in existence at war's end.

This, it seems, may have been crucial. It was then given a massive boost by the mass immigration of millions of technically-skilled workers and thousands of entrepreneurial leaders from Germany's old industrial heartlands – Lower Silesia, the Sudetenland, Saxony, Thuringia, Berlin itself – in the aftermath first of the loss of Germany's eastern territories, then of the *de facto* division of the country. And this uncannily parallels the mass migration into California in the same period. Munich, it seems, was the Los Angeles of Germany; and Los Angeles the American Munich. In both cases, against the odds, the upstart city won out against the established metropolis.

8

BUILDING TECHNO-CITIES: THE POLITICS OF TECHNO-DREAMS

Two of the most extraordinary attempts to create a techno-city for the late twentieth century are just about to take shape, in 1993, on two of the world's least likely sites. One is on a flat island in the Guadalquivir river, next to the ancient city of Seville, a major trade and cultural center of the European Renaissance, but a twentieth-century economic backwater until the 1980s. The site of the biggest World's Fair in history, the Expo '92 that took place in Seville in the summer of 1992, it was proposed as a way of reusing the Fair's infrastructures, to become the largest technocity of southern Europe: Cartuja '93, named after the island where it would be located.

The other is in the middle of a mangrove swamp, next to a soda ash works, in the city of Adelaide, capital of South Australia. Adelaide is perhaps one of the most agreeable medium-sized cities in the world, but the site at Gillman is certainly not one of its most agreeable features. But it is planned to become the location of a quite extraordinary joint Australian-Japanese venture in constructing a twenty-first-century city for living, working, and learning. And it goes by an equally extraordinary Japanese-derived name: the Multifunction Polis.

In both cases, political and business interests are battling for the control of a major project that tries to reshape the technological landscape of the city in which it is located and, with it, the city's future. Thus, at the end of our intellectual journey, the analysis of these two bold blueprints – their conception, their development, their shortcomings, and their uncertain prospects – provides an exceptional opportunity to understand the ways in which political processes shape the technological trajectories of cities and regions.

SEVILLE'S CARTUJA '93

Seville is a magic city: its glorious Renaissance monuments, recalling its time as commercial capital of the Spanish Empire at the height of its power and glory, surround the unique mystery of its Judeo-Arab medieval quarters, which are in turn built on ancient Roman foundations. It is also the city of *toreros* and *flamenco*, the city of Carmen and passion and honor, the city of art and poetry, the city of jasmine, orange trees, and fountains singing in patios of colorful mosaic and sculpted iron. Such a city – rightly proud of its two millennia of culture, of its identity, and of its superior art of living, beyond the vagaries of fashions and modernity – seems an unlikely site for the development of an advanced milieu of technological innovation in the final years of the second millennium. Yet this is precisely what the Government of Andalucía, with the support of the Spanish Government, tried to achieve here.

The Cartuja '93 Project, first conceived in November 1988, aimed at creating an agglomeration of R&D centers and training institutions, excluding all manufacturing on-site, in some of the leading late twentieth-century technologies: computer software, microelectronics, telecommunications, new materials, biotechnology, and renewable energy. And, as the project's name explicitly indicated, it planned to achieve this with a starting date of mid-1993. Yet, as soon as preliminary documents on the possibility of the project were made public, a flurry of inquiries descended on its promoters, coming from public research centers, national and regional institutions, and major multinational corporations.

By the end of 1991, a full two years in advance of the potential beginning of the project, and still with no institution actually capable of assuming its management, the number of commitments was such that over one-third of the total available space of 750,000 square meters (900,000 square yards) was already assigned, representing a concentration of no less than 1,000 researchers and about 2,000 support personnel. Already committed to locate in Cartuja '93 were six major research centers of the National Scientific Research Center (dependent on the Spanish Government), including one of the three National Microelectronics Centers and an Institute of Food Research; 10 research centers of the Andalucían Public Research System and the University of Seville; the engineering schools of the University of Seville (including telecommunications, electronics, and computer sciences); and about 20 R&D centers belonging to several major Spanish and multinational corporations, including the IBM-Artificial Intelligence Center for Voice Recognition in Spanish, Fujitsu, Alcatel, Philips, Siemens, Sony, Panasonic, Epson, CASA (Spanish Aeronautics Corporation), Sevillana-de-Electricidad, Cruzcampo (Guinness Group, focusing on biotechnology for beverage industries); ONCE (research on technologies for enabling the blind), and others.

In addition, a World Trade Center would operate a "smart building" to provide business support services, and the Andalucían Confederation of Enterprises would establish its business training center on the site. A school for advanced training in the use of new technologies, sponsored by the national Government and several firms, would locate in the Center. Last, but not least, a number of international institutions were in an advanced

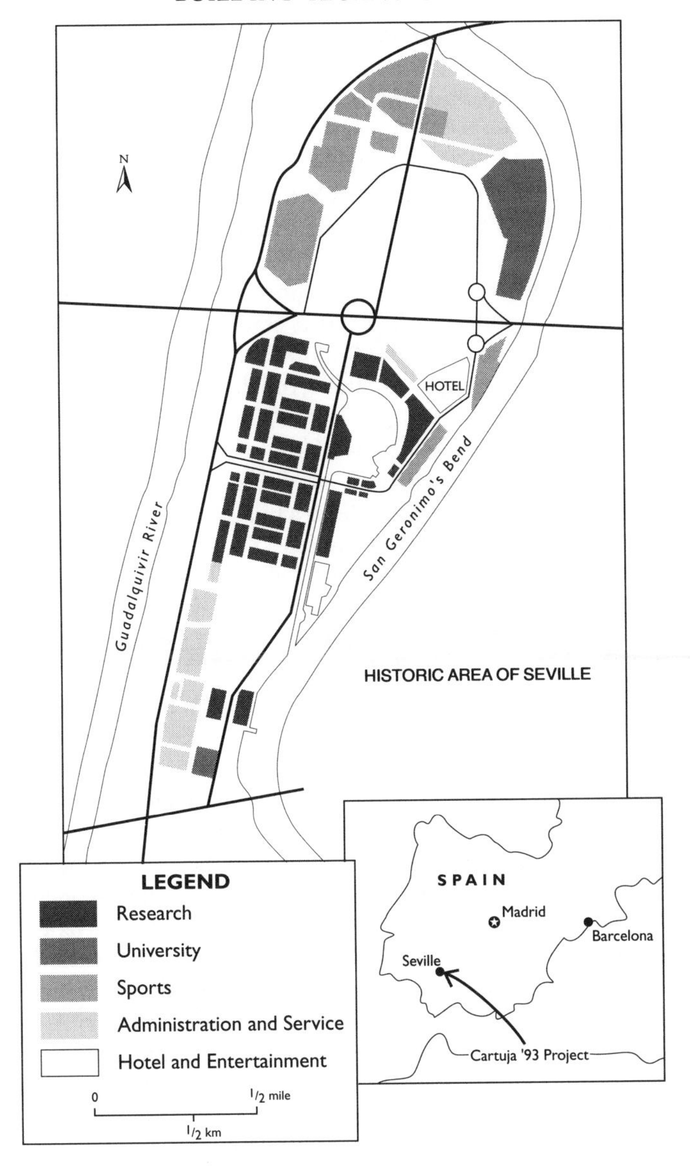

Figure 8.1 Cartuja '93: the original (1989) plan
Source: Cartuja '93: A Technological Development Project in Andalusia, 1989

stage in the process of negotiating a location in Cartuja, including the European Center for Technological Prospective, an institution of the European Community.

The projections were that by the year 2000 the Cartuja Innovation Center would house at least 50 R&D centers with about 3,000 researchers; about

30 percent of the built space would be destined to administrative operations and support services, with a total personnel estimated at some 4,000 employees; other services would include auxiliary commercial activities, services (such as banking, travel agencies, and post offices), and some leisure activities (restaurants, bars, social clubs); finally, the University of Seville's engineering campus (whose move to Cartuja is scheduled to start in 1995) would contribute 1,000 faculty members and about 6,000 students. It must be added that the Cartuja Center is located on a quasi-island on the Guadalquivir river, where, outside the technology complex itself, are already found a major metropolitan park, a sports stadium, a five-star international hotel, the National Broadcasting Building for Andalucía, and the new Administrative Headquarters of the Andalucían regional government (Figure 8.1).

But this apparent dizzying success of the project was concealing a deep controversy about the actual future of the site, with several companies arguing in favor of its transformation into a business park, thus downplaying its role as a potential technological milieu. To understand this debate, so typical of the difficulties that will very likely afflict similar future projects in other semi-industrialized regions of the world, we must first tell the surprising story of the origins of this Sevillian fantasy, thus setting the Cartuja Project in the context of the new, dynamic Andalucía, the fastest-growing region of the entire European Community in the 1986–90 period, when the project was conceived.

The Cartuja '93 project is the direct consequence of the Expo '92 World Fair, held in Seville in April–October 1992. This, the most spectacular event of its kind ever, was designed to commemorate the fifth centenary of Columbus's discovery of America. But, facing criticism that such a celebration would merely endorse an objectionable neocolonialist ideology, the Spanish Government radically changed the character of the event: it turned it into a celebration of "encounters between cultures," thus neatly putting the Spanish and Pre-Columbian civilizations on an equal level, and it redefined the fair's central theme as "The Era of Discovery." Thus, the emphasis of Expo '92 shifted from historical commemoration to the presentation of the new frontiers of science and technology on the eve of a new millennium. The new technological focus of the World Fair was crucial, for it would pave the way for the idea of a technological project as its main enduring monument.

The Spanish socialist Government, elected in 1982, inherited the 1992 Expo project from the previous centrist Government. It felt somewhat embarrassed by the idea at first sight, because of its suspect ideological undercurrents: the celebration of the Ibero-American community, a favorite theme of the Franco regime and of right-wing figures nostalgic for the long-lost colonial empire. Yet, with an opportunistic attitude typical of the Spanish socialists, the Government decided to reinforce the Expo project and to use it as a vehicle for the fulfillment of three inter-related objectives: to establish new cooperative links with Latin America, mainly based on technological and economic ties, thus shifting away from the old-fashioned culture-centered approach; to use Expo '92 as a shop window, to show the new Spain to the world; and, above all else, to seize the momentum of this project to launch a gigantic program of rebuilding the productive and communications infrastructure of Andalucía.

Significantly, both the two socialist leaders – Felipe González and Alfonso Guerra, who became respectively president and vice-president of the Spanish Government in 1982 – are from Seville, and Andalucía is the solid bastion of the socialist vote in Spain; it is the region whose unabated support for socialist policies secured the reelection of the PSOE (the Spanish Socialist Party) twice, in 1986 and 1989, besides electing the socialists with an absolute majority to the Andalucían regional government. However, it would be unfair to consider the priority given to Andalucía strictly in terms of personal and political clientelism. Andalucía was (and still is) one of the poorest regions of Western Europe, and for centuries it was oppressed and discriminated against by the local landowners and the central Government. Thus, historically, its great cultural richness went along with economic backwardness and political marginality. So it was only natural that a socialist government – coming to power after decades of authoritarian dictatorship, under which Andalucía suffered disproportionately – would make it a first priority in regional policy to bring this largest region of Spain, inhabited by some seven million people, back into the mainstream of Europe.

Using the target of the 1992 World Expo as the pretext, the Spanish Government, with the support of the European Regional Development Fund, poured public money into infrastructural projects in Seville and in Andalucía generally. Between 1985 and 1992 alone, some $10 billion (US) were spent on a variety of public works programs in Andalucía, including the building of a high-speed train linking Madrid and Seville (300 miles, 500 kilometers) in three hours, the first such project in Spain, to the great consternation of Barcelona; the construction of a whole new freeway system connecting Madrid and Seville, and Seville and the eastern Mediterranean coast through inland Andalucía, making possible the direct linkage between Seville and the European freeway network both across inland Spain and along the Mediterranean; a doubling in capacity of Seville and Málaga airports, with the construction of new, lavish terminal buildings in both; and a substantial improvement of the telecommunications system, including the digitalization of the network linking the eight provincial Andalucían capitals and the construction of a teleport near Seville, connecting the teleport with the site of Expo '92 by three optic fiber rings. In parallel, the national and regional Governments undertook a massive program of public investment in education and university research and training, that included the creation during 1988–92 of two new Telecommunications Engineering

Schools (Seville and Málaga), a new Electronics Engineering Department (Seville), and two Computer Sciences Faculties (Seville and Granada), to be added to the university potential already existing in Seville, Granada and Málaga.

The combination of public investment, new infrastructure, and the integration of Spain into the European Community in 1986, created a veritable economic boom in Andalucía. While the region had grown by less than the Spanish average for the entire century down to 1982, it has systematically grown above the national average rate between 1982 and 1992. Thus, in the period 1986–90, Andalucía grew at an average annual rate, in real terms, of about 6 percent, almost three times the average for the entire European Community, making it the fastest-growing region in Western Europe during that period. Our own study of the process of regional development in Andalucía[1] showed that it was public investment in infrastructure that made possible such growth, but also that the sources of growth included the new export-oriented agricultural activities, as well as the most technologically-advanced industrial sectors, such as aeronautics, computers, automobiles, and chemicals. In the 1985–91 period Andalucía witnessed major manufacturing investments by General Motors and Ford Electronics in Cádiz, Fujitsu and Hughes Microelectronics in Málaga, and McDonnell Douglas (via the Spanish Company CASA) in Seville and Cádiz. Together with a still strong tourist sector (albeit threatened by an emerging crisis in mass tourism), and the massive program of public works, Andalucía seemed to be on the edge of economic takeoff, combining relatively low production costs with obvious attractiveness in terms of quality of life, and access to a large, rich market through its location within the boundaries of the European Community.

The resulting image of Andalucía as "the California of Europe," irresistibly suggested by this experience, is clearly exaggerated. Yet, beyond doubt, there is genuine economic dynamism in the region; and it has been fostered, but not overwhelmed, by public investment and public support. However, our study also showed the extreme fragility of this development process: Andalucía's enduring technological backwardness makes it increasingly difficult for the region to compete in the increasingly sophisticated environment of Europe's informational economy. From this realization sprang the idea of developing a milieu of technological innovation, which would be potentially able to support the region's various productive sectors through the scientific and technological synergy that it would generate; such a milieu, it was argued, could provide the vital missing link between the process of development based on the regional investments organized around Expo '92, and the site and infrastructure of the Expo site itself. But such a link, so obvious in theory, was not so obvious in reality. As in previous case studies in this book, the history of the creation of an innovative milieu is simultaneously more accidental and more complex than the reconstructed logic of the process, as presented after the event.

The trigger for the Cartuja '93 project was the concern of the Expo '92 promoters for the future of the installations and infrastructure they were building: should they simply disappear after the magic date of October 1992, thus creating perhaps the most expensive extravaganza in history? And, if not, what could be done – and what should be done – with one of the best-equipped sites in Europe, located in the middle of one of the most beautiful cities in the world?

Indeed, the site chosen for Expo '92 was a spectacular one: on a quasi-island in the Guadalquivir River, connected by a narrow strip of land to the historic core of Seville. The island was an agricultural estate, in the middle of which stood a most original historic monument: the fifteenth-century monastery of La Cartuja de Santa María de las Cuevas, where Columbus was first buried before his transfer to his final tomb in Santo Domingo. In 1841, an English entrepreneur, Charles Pickman, whose family owned a ceramics business in Liverpool, obtained from the Spanish Government the sale of the old monastery to transform it into what became a very productive ceramics factory, making beautiful products that won several international prizes and were favored by the Spanish royal family. In the late 1980s, then, La Cartuja offered a most unusual combination: an old monastery dominated by the chimney-towers of the ceramics factory. In a gesture to historic preservation, the Spanish Government decided to rehabilitate and restore the monastery-factory – a very expensive project – to transform it into the symbol of a World Fair focusing on science and technology for the twenty-first century. It was thus that the site of the World Expo became known as "La Cartuja."

Still, this brilliant historical symbolism did nothing to resolve the problem faced by the managers of Expo '92, a Spanish public corporation commissioned by the Government to design, build, and manage the World Fair: what to do with all the infrastructure, and with the permanent buildings of the exhibition, once the Fair came to an end? There was one historical precedent: Seville had held another World Fair in 1929. The heritage had been a marvelous ensemble of buildings and parks that enhanced the architectural wealth of the city. But was this the appropriate legacy for the end of the twentieth century? Could Seville, and Andalucía, afford another round of architectural elegance, disconnected from the processes of economic development and technological modernization that were sweeping the rest of Europe? Would it not be a contradiction in terms for a World Expo, centered on science and technology, to leave as its only legacy another set of monuments and parks, connected by optic fiber?

At that point, the management of Expo '92 called Berkeley and asked the authors of this book the question they had been asking themselves. We suggested the possibility of using the advanced communications and telecommunications infrastructure of the Expo site as the starting-point for an innovative milieu that could support the development of Andalucía, as well as the process of technological cooperation between the European Community and developing countries, and using the World Fair as a launching platform for the idea. We also introduced a word of caution, recommending that first there should be a study of the process of regional development of Andalucía, which would assess the specific technological needs of the regional economy. In addition, we urged a careful analysis of the feasibility of the project before undertaking any costly initiative.

Having agreed on the methodology, we began a major study in Fall 1988, as directors of a team that included 12 professors and researchers of various disciplines from the universities of Seville, Málaga and Madrid. The project design had several built-in limitations, the most important of which was the fact that planning regulations in the City of Seville would preclude any manufacturing activity on the site; it would thus have to be a pure R&D center. An associated complication was that the Andalucían government was already committed to developing a technology park, concentrating high-technology manufacturing, in Málaga, 150 miles (220 kilometers) from Seville.

Over two years, as agreed with the Spanish authorities, we systematically diagnosed the characteristics of regional development, and the existing state of technology, in Andalucía. On that basis, we designed the project for the creation of an innovative technological milieu in Seville. We named it Cartuja '93, to make clear that its starting date could only be after the end of the World Fair, when the key infrastructural and institutional ingredients of the project would be in place. The Project was adopted by the Sociedad Estatal Expo '92 (a national government public corporation) and by the Instituto de Fomento de Andalucía (the Andalucían regional government development agency), and received strong explicit backing both from the Spanish Government and from the President of Andalucía, Manuel Chaves.

The project design, as originally formulated, included the following elements:

1 The definition of the milieu as predominantly focused on R&D activities, giving priority to those technological fields that appeared critical for Andalucían development and for the process of international north–south technology transfer. The resulting list of technologies was hardly original: it included telecommunications, computer software, microelectronics, industrial automation, new materials, biotechnology, renewable energy, food research, cultural industries, and environmental technologies. Yet we did establish the specific connection between these technologies and the productive structure of Andalucía; for instance, we related research on software and telecommunications to the new needs of the tourist industry.
2 A strong research component coming from the public sector, including six major research centers from the National Scientific Research Center, another ten centers from the Andalucían Research Program, and the

University of Seville's Engineering Schools, all of them in areas of technology designated as priority areas. The location of these public research centers on Cartuja was intended to mark irreversibly the character of the area as a scientific-technological milieu.

3 A program to attract R&D centers belonging to private firms, particularly from technology-rich multinational corporations, selecting those technologies that would fit into the overall development program. The incentives for such corporations to join Cartuja '93 were several: fiscal inducements; low-rent 40-year-leases on fully-developed land, and on the buildings themselves; excellent communications and telecommunications equipment, unsurpassed in southern Europe, including a digital switching system specially designed for the Cartuja site; public support for housing and schools for personnel; Seville's quality of life; and, last but not least, preferential agreements as suppliers of the cash-rich World Fair 1992.

4 The transfer to the Cartuja area of the engineering school of the University of Seville (4,000 students in 1992, 6,000 foreseen at the end of the century), including of course its research laboratories and training centers in telecommunications, electronics, and computer sciences. At the same time, we argued successfully to exclude the university's humanities faculties, for these would have submerged the innovative milieu under the weight of 20,000 undergraduates having little connection with technology or business.

5 The establishment of material and organizational linkages between the R&D centers in Cartuja and the productive networks of Andalucían firms, in agriculture, services (tourism), and manufacturing. These connections included easy telecommunications and communications links with the nodal centers of the region, particularly with Cádiz and Málaga, via the digital telecommunications network, freeway connections, and even helicopter service, through the location of a heliport in the Cartuja site. To improve organizational linkages, we designed a Center for Technological Diffusion and Advice, run by the Regional Development Agency, that would systematically relate the research activities conducted in Cartuja to the whole network of Andalucían enterprises, in all lines of work that were not proprietary research and thus could be sold in the open market.

6 The preliminary agreement of European and international institutions to set up research centers in Cartuja, as well as technology transfer centers that would organize the presence of Third World researchers in the Cartuja centers, as the most effective form of technology transfer. The possibility was also offered to universities of various countries to set up laboratories in Cartuja to cooperate in the research programs being conducted in the innovative milieu; in 1991, the University of Puerto Rico had already decided to establish one such center.

7 The management of the innovative milieu should be in the hands of a

> largely autonomous public corporation modeled on the British New Town Development Corporations, with the participation of the regional government (in our view, the dominant authority), the Spanish Government, and the City of Seville. Private firms present in Cartuja would be associated and represented in the decision-making body. In October 1991, such a public corporation, termed Sociedad Estatal Cartuja '93, was indeed established, under the majority control of the national Government, with representation of the regional and local governments.

This well-crafted design, distilled from international experience, soon became all-too-predictably compromised by the complexities of politics and by narrow, short-term speculative considerations. Indeed, long before we had completed our study and designed an ideal milieu of innovation for Andalucía, the word was out and inquiries were pouring in: at first sight, a testimony to the commercial soundness of the project. But in fact, the main attraction of the project lay in its access to relatively cheap prime land, in a symbolic site, with unsurpassed communications equipment, in the highly attractive capital city of the most dynamic region of the new Europe. Real estate entrepreneurs, organized around the Spanish World Trade Center, set up an Association of Users of Cartuja '93, constituting the major private firms involved in the project, and persuaded them of the value of transforming their R&D centers into prime office space at a premium price; subtle cosmetic treatment would allow them to represent as R&D what was in reality, with few exceptions, a series of company headquarters offices for southern Spain and the Mediterranean area. When the Andalucían government, following our advice, proposed the creation of a Unit of Technological Evaluation to screen the applicants, the chair of the Association, himself the president of the World Trade Center, an office-building operation, led an attack on the entire Cartuja '93 project, precipitating a crisis between the private sector and the public interest.

The crisis even split the public sector representatives themselves: it opposed the Sociedad Estatal Expo '92 and the Andalucían government, both controlled by the Socialist Party. For the managers of Expo '92 suddenly realized that it would be relatively easy to make quick profits by selling entry rights to Cartuja without stressing the R&D requirement; as well as showing the new Spain to the world for six months, Expo might now even recoup its huge costs, a goal much more valuable to the management than an uncertain bid for the technological future of Andalucía. On the other side, the Andalucían government continued to support the technological project: even though the initial private-sector response might well be slower, because of the greater risk, the long-term impacts for the region's economy could be huge.

The debate on the future of Cartuja '93 intensified a few months before the opening of the World Expo; as a result, the Expo '92 managers disengaged from the process, to concentrate on their priority task of assuring the success of the World Fair. The Madrid Government appointed a top civil servant as president of a new public corporation, Sociedad Cartuja '93, and gave him the virtually impossible task of somehow putting everything together into a package. And, to complete a quintessential Andalucían imbroglio, in May 1991 the socialists for the first time lost the Seville municipal election. The new mayor, the leader of a center-right regionalist party, decided to make his own mark on the Cartuja '93 project, favoring the real estate and business interests that had supported his electoral campaign.

Thus, he argued for extra space to be allocated for commercial activities and office development, and for a "flexible" criterion to evaluate the technological content of the R&D centers locating in the science and technology city. In addition, the conservative mayor of Seville, a traditional *señorito* named Rojas-Marcos, was keen to show his independence from the socialist Government by pushing an alternative project for the reutilization of the World's Fair installations: the building of a techno-cultural park that would emphasize amusement and leisure over research and training. The techno-cultural park would include the preservation of some of the museums and exhibition halls of the 1992 World Fair, and would encourage the presence of bars, restaurants, and other leisure activities on the Cartuja Island. This proposal was well-received by the local population because it could quickly provide a large number of badly-needed unskilled jobs bringing some relief for rising unemployment in the construction industry, once the major public works for the World's Fair were finished. In addition, by linking up with the Sevillian tradition of feasts, celebrations, and amusement, the project blended better into traditional Andalucían culture than the bold attempt to propel Andalucía into a path of twenty-first century, technology-led economic development.

However, the City of Seville held only one out of 10 seats on the board of the public corporation managing the Cartuja 93 project. The ambitions of the mayor seemed doomed to failure, against the potential opposition of the five seats controlled by the national Government (socialist), and the four seats controlled by the regional government (also socialist). And yet, in the end it was the blueprint of the mayor of Seville that appeared to gain the upper hand only a few months before the opening of the Cartuja '93 park. In April 1992, the public corporation board approved a new master plan for Cartuja '93 that assigned the land users in roughly equal proportions (one-third for each use) to office-commercial development, to a techno-cultural park, and to R&D centers and university facilities. In addition, private R&D centers were allowed to include a substantial proportion of non-research office space, actually reducing the overall R&D utilization of the complex to about 25 percent of the surface area. Thus, while some R&D activities will still be concentrated on Cartuja (making it the main R&D

center of southern Spain), the original project aiming to create a milieu of innovation as the core of the new development process of Andalucía was for all practical purposes abandoned.

The seemingly unlikely victory of the mayor of Seville can be easily explained by a combination of factors, all of them full of lessons for future similar projects in other areas of the world.

First the mayor effectively derailed the whole process, claiming for the city the control of land-use planning prerogatives that the regional government had assumed for itself. Although the legality of the regional government's move could have stood in the courts, the claim by the City of Seville, if presented, would automatically paralyze the process of building and developing the science city for at least three years, losing momentum at the critical time in the aftermath of the World Fair, and opening a very dangerous transition period during which the environment would have deteriorated, and investors would have withdrawn their commitment.

The regional government was particularly sensitive to such a threat, since the potential unemployment problems in Seville could be aggravated, while the most valuable space inherited from the 1992 celebration would remain idle, exposing the government to political criticism. Of course, such blame could also be laid at the door of the mayor of Seville. But in a political battle, it is always the most determined (or the least responsible) who wins.

In addition, the regional government had another problem that undermined its support for Seville's milieu of innovation: the crisis of the technology park in Málaga. This park was designed by the regional government in 1989, around the location of a Fujitsu computer factory and, later on, of Hughes Microelectronics. Problems in the development of the park, and yet another political rivalry between the mayor of Málaga (a socialist), the University, and the regional government, paralyzed the technology park to the point of jeopardizing its existence. Since the national Government favored the Seville project, giving to it substantial fiscal incentives, Málaga's political elites put such pressure on the Andalucían government that it had to concentrate its own regional incentives on the Málaga technology park in an effort to save it.

Nor was the national Government ready to fight to the end for the technological future of Andalusia. Since the inception of the Cartuja '93 project, the economic environment had turned sour. Spain was no longer in the middle of an economic boom. The budgets for 1992 had been vastly overspent (in Andalucía and Barcelona alone, Spain spent close to $15 billion in infrastructure, installations, and management in the 1985–92 period), and the moment had come to obtain as big a return as possible in order to alleviate the public deficit. Thus, the faster the selling or leasing of the Cartuja '93 project, the better the accounts of the public corporations managers would appear to the eyes of the Ministry of Economy.

Finally, Andalucían business used all its political and personal influence to be able to locate in the best equipped urban space of southern Europe. Given the short-term, speculative character of most of these companies, they could not care less about the long-term interests of regional technological development, if such a perspective would not pay off for them in the very short term. Also, as soon as multinational corporations involved in the project understood the situation, they too tried to take advantage of the new opportunity, maximizing their office space at the expense of R&D facilities.

Confronted by all these obstacles, the public corporation tried desperately to balance the different interests and projects, by allowing the location of all the various activities claiming the site: amusement and recreation, techno-cultural programs, shopping areas, offices (both public and private), the engineering schools, public research centers (but not all those originally planned), and some private R&D centers blended with office space for the same companies. The last stand of the "Cartuja '93 Public Corporation" was to reserve the considerable fiscal incentives accorded by the Government to R&D activities, so that the locational benefits already enjoyed by the companies locating in Cartuja 93 would not extend to a tax-break under the false pretext of being part of a "technology park."

Doubtless fascinating in academic terms, this high-tech soap opera nevertheless has potentially significant consequences for the future of Andalucía. The authors would have liked to wait for the denouement, since we are convinced that the final episode is not complete. But since this book must be finished, so as to bring its lessons to other regions around the world, we close the chapter in this unfinished state – with two final caveats:

1 In spite of all the contradictions and conflicts we have described, and in spite of the abandonment of the original project of designing and developing a milieu of innovation, a concentration of scientific research and technological innovation is being built in the middle of a city that until recently, despite its history as a center of culture, was simply a technological wasteland. To be sure, business pressures and political interests will create some hybrid of an office complex, a techno-amusement park, a university campus, and an R&D center. Nevertheless, Seville could be in the process of emerging on the European map of science and technology. For instance, there is a serious possibility that Seville will become the second CERN (European Nuclear Research Center) for advanced high-energy particle research in Europe, with the building of a Tau-machine particle accelerator, and the establishment of a permanent research center housing over 100 top physicists from around the world. Although a final decision on this project had not been taken at the time of writing, the fact that it was under consideration is significant evidence for our original proposition concerning the potential of new regions for scientific-technological innovation, provided the right mix of elements goes in at the right time.

2 Finally, it is well-known that the political environment is critical for the fate of all development projects. But this is even more true in the case of an ambitious project that goes against spontaneous locational market forces and the fact of technological inertia. Thus, unless government institutions are firmly behind such a technological development project, the mediating forces and interests that inevitably arise along the way will divert it from its original goals. How much these goals will be unfulfilled, as a result, will depend of course on the degree of the deviation, which in turn will be a function of a complex set of relation ships between speculative interests, business development strategies, and the political vision of the public institutions.

Whatever the outcome, Cartuja '93 will undoubtedly be a part of the new reality of Andalucía. But at the time of writing it is still unclear if it will merely add another chapter to a long and unhappy local tradition, whereby a few *señoritos* appropriate scarce wealth for their personal advantage; or if it will, after all, signal a new beginning for the region, uniquely marrying the capacities for leading-edge production and the arts of gracious living.

ADELAIDE'S MULTIFUNCTION POLIS

The MFP, as Australians almost immediately initialized it, is Australian in location but Japanese in name, Japanese in origin, and, perhaps, peculiarly Japanese in its entire basic concept. It made its first appearance at an Australia–Japan Ministerial Committee meeting in Canberra in January 1987, where the key players were Hajima Tamura, then Japan's Minister for International Trade, and Australia's Senator John Button, then head of the corresponding ministry, DITAC (Department of International Trade and Commerce). After that meeting, a release described "a multifunctional facility . . . [which] would incorporate future-oriented high technology and leisure facilities and could promote international exchange in the Pacific Region on new industry and lifestyle."[2] Some Australian DITAC officials have believed ever since that the Japanese side developed the idea as the equivalent of the gift (*omiyage*) they take on important social occasions, and were surprised that the Australians took it so seriously.[3]

Whatever the degree of seriousness, the MFP had started a few months before as yet another brainwave in the collective mind of MITI, Japan's Ministry of International Trade and Industry, and its ancestry clearly stems in part from MITI's earlier technopolis project. But it had another, equally interesting, origin: in MITI's 1986 "Silver Columbus" plan to export older Japanese to retirement colonies abroad, in order to relieve population pressure. Reviled in Japan, this scheme was quickly abandoned; but, significantly, the MFP originated one year later, in the same (Leisure) division of MITI. And it could be said that the two ideas – technopolis and retirement – here came together, for the technopolis as well as the ideal retirement colony were now to be exported in one package. An early MITI draft on the MFP, produced only a month after the Canberra meeting in February 1987 and never made public, describes it as "an International, Futuristic and High-Tech Resort through Australia–Japan Co-operation" and as a "cosmopolis to become a forum for international exchange in the region and a model for new industries and new lifestyles looking ahead to the twenty-first century"; it was to combine a high-technology function ("as a model for new industries"), a resort function including recurrent adult education ("as a model for new lifestyles"), and a cosmopolis function ("as an opportunity for exchange").[4] Contrary to rumors that circulated later, it was specifically designed for Australia; it would build on the coming bicentenary, the opening of the National Science Centre in Canberra (funded partly by Japan) and the 1988 Brisbane World Expo, and it even appears that MITI had already identified Brisbane as the most promising site.[5]

By September 1987, MITI had produced a much longer (65 pages against 11) and more elaborate "Basic Concept," a pre-feasibility study that included some very exotic notions; so exotic, indeed, that they were soon pilloried by Australian critics. One was the "City of the Fifth Sphere" which was supposed to combine all previous historical stages of urban evolution: in the first, home and workplace were combined; in the second, they were separated; in the third, recreation emerged as a separate realm; in the fourth, extended-stay resorts and a diversified lifestyle were said to develop. One evidently sceptical Australian critic comments: "The notion of human history implied in such a scheme, was charming in its naive idiocy; the combination of one and two was even a logical impossibility, while four was clearly the Japanese domestic 'resort' strategy in its export form."[6]

Be that as it may, the MITI concept involved the creation of a new kind of milieu combining work, leisure, health, and education within a single new city of about 100,000 people, located somewhere in Australia. High-technology industries, already targeted by MITI as the leading-edge technologies of the early twenty-first century – computer and information technology, biotechnology and health sciences, new and rare materials – were to be joined by "high-touch" industries like tourism, entertainment, and sport in a new kind of urban cocktail[7] within a human-scale, semi-residential city (i.e. a place where people stayed for periods between a few weeks and several years);[8] this, in turn, would require formation of a "software infrastructure," consisting of diversified information and leisure systems representing a "completely new structure that has no precedents today and therefore is expected to generate new industries."[9] In a phrase that soon acquired some notoriety in Australia, MITI encapsulated its vision:

> The MFP is a place of providing, gathering, and reproducing information of diverse aspects, strata, and form, as well as for relaxation, comfort, surprise, joy, entertainment and intellectual stimulation.[10]

An Australian commentator later somewhat acidly remarked: "Precisely what the MFP might be and how much information might be combined with how much joy and intellectual stimulation, remained to be worked out."[11]

One evident fact about the concept is that the "high-touch" or resort-convention element was more fully developed than the high-tech element. According to the February report, it would include sports and health facilities, fine arts facilities, a school of foreign cultures, and a business school; the amenities would include condominiums owned by corporations to house employees attending refresher courses. In the September report, Honolulu is mentioned as the closest model.[12] For this emphasis, there were good Japanese reasons. Modern Japanese society is notoriously characterized by inadequate leisure and high stress, but until recently the only reaction seems to have been ever more bizarre commercial schemes such as "100 PSY Brain Mind Gym Relaxation Salons" and womb-like "Refresh Capsules." In 1987, however, came a more sophisticated government reaction in the form of a Comprehensive Resort Region Provision Law, which stimulated competition among localities to such a degree that by December 1989 19.2 percent of the entire land area of Japan, twice the area devoted to agriculture, was involved: 646 projects were at the "works" stage and another 205 at the planning level, rejoicing in such suggestive names as Mie Sunbelt Zone, Aizu Fresh Resort, Snow and Green My Life Resort Niigata, Gunna Refresh My Life Resort, Chichibu Resort, 40° Longitude (sic) Seasonal Resort Akita, and Nagasaki Exotic Resort.[13] At one blow the law neatly provided an answer to the critics in agriculture and forestry resulting from trade liberalization; encouraged regional development; served the politically important construction and real estate lobbies, and offered the promise (possibly a hollow one) of a more leisurely lifestyle for the average person.[14]

But the law did even more: it recognized that, throughout the 1980s, more and more capital had been flowing out of Japan into tourist-based real estate developments abroad. By the end of the decade, many of the first-class hotels, golf courses, and luxury hotels and apartment buildings in places like the Gold Coast and Hawaii were in Japanese hands; Japanese funds flowed into the Gold Coast at the rate of over $A1 billion a year in the 1980s, rising in 1990 to $A7.25 billion in Surfers' Paradise alone.[15] In 1979 80,000 Japanese tourists visited Australia, in 1989 350,000; the number is expected to exceed one million by the century's end. Australia has been implanted on the Japanese popular imagination, mainly through media images.[16]

The underlying strategy of the MFP

This describes the MFP as officially presented. The critical question, over which Australians have been agonizing ever since, is what reasoning underlay it.

Some conspiracy theorists have seen the MFP as invented by Japanese corporations with a big hand in the Australian economy, like Kumagai Gumi and EIE. Japan's economy, as these critics are not slow to point out, has been increasingly based on what could be called a political-construction complex, which has been outstandingly successful in what the Japanese call *doken kokka*: diverting a steadily-increasing volume of public funds, derived from the apparently inexorable rise in Japanese tax rates during the 1970s and 1980s, into huge regional and urban construction projects such as bridges, tunnels, highways, high-speed railroads and airports, and latterly into huge resort complexes.[17] This formula seems to have been successful enough, so far, in Japan: until the 1993 election, it spread enough benefits to the regions to support the ruling LDP's power base, while providing contributions from construction and related industries to maintain the party's campaign funds. The MFP, it can be argued, appears to be intended to extend this concept offshore.

Shugo Minagawa of Nanzan University argues however that the concept is too idealistic and naive for that; it came from deep within MITI, and there is some basis for thinking it was a partly-reworked version of the "Silver Columbus" scheme.[18] Certainly, MITI had an urgent motive of its own for marrying the technopolis and resort concepts: the fear of being demoted to inferior Cabinet rank because of the shift of the Japanese economy from export-oriented to demand-driven. By the late 1980s, MITI was battling with Transport, Construction, and Posts and Telecommunications for second position after Finance; this, it may have been thought, would help counter negative publicity for the "Silver Columbus" proposal via a new concept combining high-tech, resort, and life education.[19]

It seems plausible, in fact, that for MITI the MFP offered three separate advantages, which would help advance the Ministry's status within the Japanese governmental pecking-order. First, it would help improve Japan's image and public relations in a new international venture, centered on Australia but extending across the Pacific. Second, it could involve Japanese companies in international exchange of R&D information. And third, it could help generate creativity.[20]

A new international venture

The first reason was concerned with high international politics: to counter North American and European attempts to create free-trade zones, Japan needed to sponsor an "Asian–Pacific" zone headed by Japan.[21] The prime

Japanese reason for choosing Australia for this enterprise was that it offered a unique opportunity for a truly joint venture and learning experience; Australia had the land and the resources, Japan the capital and the know-how, and both are in the same time zone, facilitating communication. MITI saw Australia as an environment where information-intensive industries could be more fully developed than in Japan, as well as exposing large numbers of Japanese people and businesses to international influence, at the same time encouraging them to change their lifestyles, and place greater emphasis on leisure and improved living environment.[22] And Australia was attractive for Japanese investment because of political stability, a high level of education and skills, proximity to Japan, and solid infrastructure.[23]

R&D exchange

The exchange of R&D would come about because the pattern in the two countries is quite different. In Japan it comes from private industry: indeed, Ian Inkster argues that "the 'gap' between Japanese universities and Japanese industry is as great, possibly greater, than that of other, major industrial nations."[24] The private sector relies on advanced training based on rotation within the firm and employee-sponsored correspondence courses, the costs of which are paid by the firm; the stress is on specificity, flexibility and efficiency rather than general "education."[25] Australia in contrast is characterized by very high government spending on R&D, especially through CSIRO (the Commonwealth Scientific and Industrial Research Organization), and by very weak private participation: the private sector raises only 37 percent of R&D funds compared with 68 percent in Japan, 65 percent in Germany and 47 percent even in the USA.[26] While 56 percent of Japanese small firms (with 300 or less employees) conduct some research, the corresponding Australian figure is 3 percent; Australian firms largely lack the less sophisticated applied research that allows smaller companies to keep in touch with trends and adapt innovations to their own needs, which has been so widespread in Japan.[27] So, the implied argument runs, an exchange between these two very different traditions could be mutually advantageous.

Creativity

The notorious "relaxation, comfort, surprise, joy" passage in the Basic Concept needs to be interpreted as a response to the debate that has been raging within Japan about "creativity": Japanese institutions, group norms and procedures, it is argued, fail to generate creative, innovative ideas and fundamentally new technologies; therefore, the MFP should act as an institutional innovation to foster creativity. Spontaneity and "joy," which in some way are connected to creativity, will be generated within the MFP and encouraged by day-to-day contact with Australian and other foreign scientists, engineers and officials, all generated in the high-touch environment of conferencing, information services and education. In this interpretation, the "pleasure dome" element is profitable both in itself and as a service to high-tech creativity.[28] Against this, Tessa Morris-Suzuki argues that the problem is not one of creativity, but of demand: recent advances in information technology have generated relatively little in the way of fundamentally new consumer products to rival the consumer revolution of the mid-twentieth century, and in Japan the problem is exacerbated by the acute lack of space in the average Japanese home.[29] Whichever the case, there could be no doubt that senior MITI officials saw the need to generate a completely new range of basic innovations, equivalent to those that had created the great consumer revolution of the period 1950–90.

The chronology of the MFP

Public announcement of the feasibility study followed in January 1988; a Joint Steering Committee was appointed in July 1988 and consultants, Andersen-Kinhill, were appointed in November 1988. Their pilot study was finished in March 1989 and their final report in December 1989.[30] Soon, a quite extraordinary conglomeration of major companies lined up on both sides: 86 in Japan, and 62 on the Australian side.[31]

These two-and-a-half years of frenetic activity were however marred by a quite extraordinary degree of chaotic management and mutual misunderstanding, and by sometimes vituperative criticism on the Australian side. For this, the vagueness of the original concept might be partly blamed; equally, it seems true that both sides were interpreting it the way they wished, without much reference to the other. As one Australian commentator summed it up in 1991:

> By any normal reckoning, the Multifunction Polis probably should have been dead and buried by now. It has suffered at the hands of a generally sceptical and ill-informed media, it was dragged through a divisive and bitter Federal election campaign, and it has been attacked in public meetings by critics on the Left and the Right. Worst of all, those charged with conducting the Feasibility Study into the proposal seriously mishandled their task: they failed to undertake effective public consultation; they sent the study running off the trails into wild areas of speculation (notably the possibility of an 80 percent foreign population); and then they allowed the final site selection process to turn into a public "catfight". Yes, the fact that the MFP has survived what former South Australian premier John Bannon called, with numbing understatement, its "bad start" is a marvel in itself. That is not to say that the damage of three rough years won't prove fatal in the end.[32]

A first basic problem is that, as seen, MITI were thinking of a "high-touch," resort-led type city, while the Australians were expecting high-tech: a rich recipe for misunderstanding.[33] A second was that, from the start, the Australians insisted on seeing this as essentially a private-led scheme, which was not to be financed through location-specific Commonwealth and state subsidies.[34] This was inevitable, because there was no Australian policy framework within which the larger ideas could be tested.[35] As Hamilton has argued, this probably contradicted the entire Japanese approach to schemes like these, which was based on:

> transcending short-term opportunity, teasing narrow interest into a fully developed profit and non-profit enterprise of high visibility for Australians and the international community. It means having a place and a plan, and both require taking a dare on the future. Only governments, in full consultation with their electorates, are in a position to make that kind of commitment.[36]

A third was that the two sides were divided on the question of priorities: the Japanese wanted a site quickly designated and the character and components settled later, while the Australians wanted first to decide on the "multi-functions" before turning to the location of the "polis."[37] And fourth, suspicions multiplied in Australia because of a technocratic and manipulative approach to public opinion.[38] In particular, the rumor that the scheme might involve the settlement of some 200,000 Japanese caused many Australians to balk, because they could not be expected to appreciate that the great majority would be short-term (perhaps 30 percent on three-month contracts, 70 percent on three-month to three-year contracts), the economic effect of which would be analogous to tourism rather than migration.[39] "The pathetic notion that Japan wants to 'colonise' Australia by creating an exclusive, elite city danced like a will-o'-the-wisp through the public debate."[40] No one in Australia seemed to realise that the "fifth sphere" came out of the uniquely Japanese urban experience of overcrowding and inconvenience.[41]

Finally, and almost inevitably, there were epic arguments over the choice of location. The 16 criteria laid down in the Spatial Attributes Analysis, commissioned by the Joint Steering Committee and published in July 1989, strongly suggested that the right location would be in and around one of Australia's major cities; indeed, they pointed strongly toward Sydney or Melbourne and to the hub-and-spoke model suggested by several agencies, including New South Wales.[42] But the Japanese insisted on a single site, and this may have been the final element in the elimination of both Sydney and Melbourne from the contest in late 1989.[43] An economic evaluation, produced by the Bureau of Industry Economics in Canberra in 1990, suggested that only Sydney and Melbourne have the necessary concentration of research scientists in and around them.[44]

There was yet a further complication. For the first three years of the MFP's gestation, it was closely associated in the Australian public mind with another huge project: the Very Fast Train (VFT), which was to run at 220 m.p.h. (350 km/hr) on a new, dedicated track over the 320 miles (550 kilometers) between Sydney and Melbourne via Canberra. The notion was that a new city – or even, in one incarnation, several new cities – would be built along the route of the new line, which would connect with international airports and thus establish the global character of the MFP, at the same time helping bolster the financial viability of the rail project. Finally, soon after the decision to locate the MFP in Adelaide – 800 kilometers (500 miles) away from the new rail line – a big argument developed regarding the viability of the VFT: as a privately-financed project (albeit with Commonwealth backing), either it would require land concessions or tax concessions to make it pay. Finally, getting neither, it was put on ice in late 1991.

Eventually, on June 1 1990, came a decision: it was in favor of a site on the Gold Coast of Queensland. The award was supposed to be announced on Friday, June 15, 1990, but after Queensland premier Wayne Goss was awakened at 3 a.m. in his Seoul hotel and told he had to deliver an immediate land purchase package, the deal collapsed, and a week later, on June 19, the decision went to Adelaide.[45] In fact, the award was compromised from the start by the Joint Steering Committee's insistence that the state government should appropriate the increase in land values that would result by immediately buying up the land and freezing the process.[46] The land in question was 4,700 hectares (11,500 acres) of prime real estate between the Camera and Pampama Rivers, with frontages into the Broadwater behind South Stradbroke Island; most was independently the subject of proposals from local and overseas interests, including Japanese firms, and principals in the 2020 Syndicate that organized the MFP bid; 757 hectares (1,870 acres) had just been bought by Hokojitsugyo, a little-known Japanese firm, for a reported $A66.8 million. But some land was divided up among many small local owners, who were understandably opposed; and the resulting political fight caused Queensland to back out.[47] Wayne Goss, the premier, gave up and asked the Commonwealth Government to take over the site selection process; at that point, the decision came in favor of Adelaide.

Adelaide had always been second-best in both the first and second rounds; but it had never been far behind,[48] and its bid stressed economy and practicability: the extra 7,000 people per year represented a 50 percent increase on 14,000, easily manageable; there was no need for a new airport or a very fast train, and the approach to the World University would save nearly $A1 billion. The costs were supposed to be shared: South Australia $A0.2 billion, the rest of Australia $A1.0 billion, "other" $A4.8 billion; Hamilton describes this as an "imaginative calculation."[49] Some argue that the South Australian site is not optimal, because it is less likely than Sydney or Melbourne to fulfill the functions of high-tech production, pre-commercial

R&D and information dispersal.[50] Nevertheless, the choice of Adelaide was endorsed in the final report of the Joint Steering Committee in September 1990.[51]

MFP-Adelaide

In short order, an MFP-Adelaide Management Board came into being; it ordered yet another feasibility study, this time of the Adelaide site. It reported in May 1991; predictably, the report was favorable. At the end of July 1991, the Federal Government formally approved the project, with an allocation of $A12 million for start-up costs, including $A5.5 million to the South Australian government to help South Australia meet the costs of marketing and of establishing an MFP Development Corporation; the plan was to start work on the site in late 1992 or early 1993.[52]

The Adelaide project is thus very different from the original MFP concept. Far from being a completely new twenty-first-century demonstration city, it is to be sited on the only large piece of open land (4,550 acres, 1,840 hectares) remaining within a 15-mile (25-kilometer) radius of central Adelaide. Only five miles (eight kilometers) away from the city center, it has remained undeveloped until now because of the extreme difficulty of preparing the land. Like other hillside cities worldwide, almost from the start Adelaide grew up the cooler hill slopes, leaving the flatlands near the sea for port and heavy industrial uses. So, on its west side, the MFP Gillman site faces the old Port Adelaide, now partially restored as a historic landmark, plus a row of heavy industrial plants. It is rather as if Silicon Valley was to be relaunched at Richmond on the east side of the San Francisco Bay, next to the oil refineries. From here, the site extends as a series of smaller discrete developments up the Lefevre Peninsula, to the northwest of the main site, through Port Adelaide to Outer Harbour at the very tip (Figure 8.2).

Nevertheless, despite the abrupt change of locale, the Adelaide MFP contains many of the elements of the original scheme. According to the May 1991 report, it is to be "an Australian city with a world vision." It will be, says the report on its opening page:

1 An international centre of innovation and excellence in urban development and in advanced technology to serve the community;
2 A leading centre of innovation in science and technology, education, and the arts;
3 A national focus for economic and technological developments of international significance;
4 A model of conservation and management of resources and the natural environment;
5 A focus for investment in international business development based on new and emerging technologies that will form the basis of the economics of the twenty-first century;

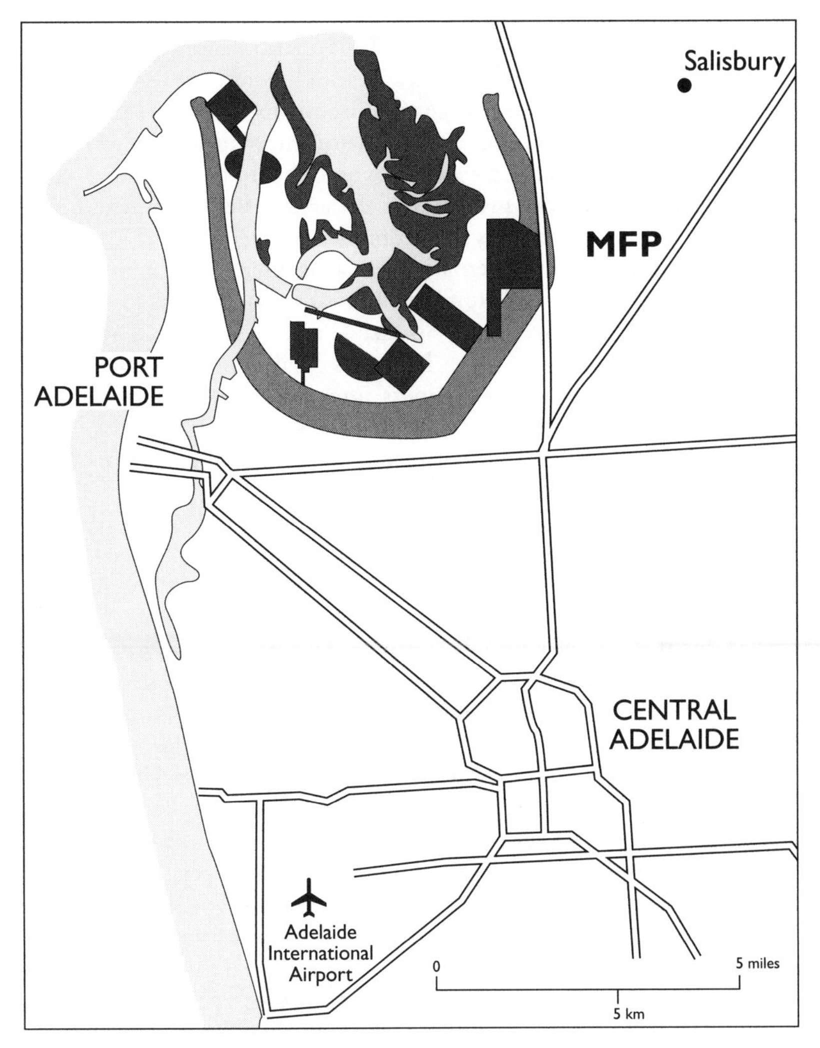

Figure 8.2 Adelaide Multifunction Polis: general location
Source: Adelaide: The Nominated Site for the Multi Function Polis (n.d.)

6 A social model for the twenty-first century based on equitable social and economic development.[53]

Specifically, the MFP would feature centers of excellence in information technology and telecommunications, advanced learning technology, environmental management, health and urban and community development, forming the core of a national and international network lining such centers; and an Australian center for the management of innovation, "particularly in the cross-overs between technology, environment, society and business"; a major channel for commercializing and internationalizing the outcomes of Australia's scientific and technological research; and an international forum for research, information exchange and discussion on key areas of endeavor and, in particular, joint approaches to major international issues.[54]

Even more specifically, the MFP would be built around a series of research and development institutes in the areas of environmental management, information technology, telecommunications networking, advanced learning technology, and international management and innovation. Each institute would have a separate focus, but all would be linked as structural elements of "MFP University," an Asia–Pacific institute of research and higher education. These research institutes "will work closely with MFP-Adelaide investors and other business interests to commercialize leading-edge research and development results."[55]

The MFP University would be a new institution, focusing on "research carried out in a series of institutes built up to world-class standard by collaboration between strong elements in existing South Australian universities; the addition of new, complementary elements; and the development of effective links with institutions and industry in Australia, the Asia–Pacific region and other parts of the world."[56] It would encompass a Learning Systems Institute, including an Advanced Systems Research Centre and a Distributed Education Service; an International Management Centre; an Environmental Management Research Institute; an Information Technology and Telecommunications Institute; an Asia–Pacific Institute of Language and Culture; and a Health Research Institute. Each of these, it is made fairly clear, would be a heavily applied research center, intended to produce a stream of commercial applications including products and services. Somewhat as an afterthought, in a subsequent chapter the report mentions that "the development of a space industry focus for MFP-Adelaide will offer a number of benefits," though this had not been discussed in the chapter on research.

Information technology and telecommunications are a key element of the concept. The report claims that by the early twenty-first century, MFP will become a city with advanced telecommunications systems and services; a key site within Australia for the location of software and services firms that will serve the Asia–Pacific markets and the demands of global firms; an engineering center of Asia–Pacific regional significance; and an Asia–Pacific center for the trial of prototype IT products, particularly those used by the individual and the home, including especially automatic translation projects to bridge the Asian–English language gap.[57] To this end, the MFP is built upon a proposal by the South Australian Government to establish an information utility offering a network for integrated voice, data, and radio services coupled to value-added services, a common interface for public and private users, and large-scale general computing services.[58]

MFP-Adelaide would be more than just a technopark on a huge scale, however; when complete, after a 20- to 30-year development period, it would also be a major town in itself, with some 40,000–50,000 residents living in a series of small mixed-use villages (thus minimizing the need to commute), 2,000 by 2,000 feet (600 by 600 meters) in size and housing 2,000–2,500 people, separated from each other by lakes and forests; 60 percent of each village would be residential, 30 percent would be in community, commercial and mixed use, and 10 percent would be set aside for public space. The first stage, comprising between three and five villages, would house some 10,000 people over a five-year period and would cost $A120 million for the land development with an additional 30 percent government contribution and another $A200–250 million of housing costs borne by the developer.

The MFP-Adelaide report is, to say the least, an intriguing document. Though it is supposed to be a feasibility report, its analysis of feasibility is contained in a single statement that a project internal rate of return of approximately 24.3 percent is achievable – a statement nowhere supported by further evidence in the report itself. The report does admit that "From both an engineering and a marketing perspective, the site will require significant initial investment in the open space and lake system."[59] Infrastructure investment costs are calculated at $105 million, equivalent to a continuing additional annual cost to the state government of only $9 million in 1991 dollars. This investment, it is estimated, could meet up to 15 percent of Adelaide's demand for urban land, thus deferring or avoiding costs in bringing services to fringe areas; this argument, too, is not developed in detail. The report concludes that "MFP-Adelaide needs to succeed on its own merits without project-specific government incentives," but suggests that a number of existing government incentives, both at state and Commonwealth level, should be of interest. These include possible concessions in state and local taxes and rates, to be negotiated on a case-by-case basis, and the acquisition by the state of minority equity in projects of major significance.

Conclusion: the MFP in 1993

In the Fall of 1992, South Australia acquired a new government; the new administration however maintained continuity with the old, by moving quickly to introduce the legislation to set up the MFP Board. By early 1993, the Board was in place and the appointment of a high-profile chief executive was said to be imminent. An Environmental Impact Report had been published, showing – perhaps unsurprisingly – that the site could be cleaned up if enough money was spent on the job. One good feature emerged in the form of an ambitious scheme to clean up the site, with innovative ideas for storing and reusing storm water and making use of industrial waste products. Sixteen private companies had announced plans to set up an integrated services company to provide water, gas and electricity to the MFP and to handle the disposal of domestic and industrial waste and sewage. In the plan, waste from one activity would become a productive material for use in another; by functioning as an integrated company, the consortium would provide a demonstration project of conserving natural resources and minimizing waste, thus developing technologies that could be exported elsewhere. The company, which is due to be launched in late 1993 or early 1994, is claimed to be the most highly-integrated waste management project in the world.

One conclusion is compelling: although the locale of the Multifunction Polis may have moved a long way, the essence of the scheme remains essentially unchanged from the original MITI proposal. This is evident even though, remarkably, Japanese participation is never once mentioned in the Feasibility Report. The essence, difficult as it is to grasp from the very generalized and even vague language of the report, appears to be this: MITI's innovation policy increasingly stresses the need to move upstream into more basic creativity, by developing the links between basic research and industrial application. But, so far in Japan itself, these have proved bafflingly elusive: the Silicon Valley model of synergy has not been achieved. Perhaps, the implicit argument runs, this can be achieved only in a non-Japanese context; or, more precisely, by creating a new kind of Co-Prosperity Sphere in a Western country. Australia, which has a Western culture including a British-derived university system, is an ideal location, partly because – in comparison with the North American continent – the universities have not yet achieved full benefits of scale and could benefit from infusion of research funds. Indeed, it is difficult to resist the idea that MITI may have identified the Adelaide region as the San Francisco Bay Area of the early twenty-first century: the source of the next great innovative long wave of history, this time under Japanese auspices.

The MFP, therefore, is a logical extension of the strategy MITI and other government agencies have followed in Japan during the 1980s in the form of technopoles, teletopia and similar programs. Further, it offers scope for the Japanese construction industry, which has specialized in huge waterfront land-reclamation schemes, to find a major overseas outlet in cooperation with its well-established Australian counterpart. And this could be significant to the Japanese, because there must be a limit to such schemes in the overheated, overcongested, speculative Japanese urban land market – even with the new stress on locally-based resort developments.

Such a logic, if it is indeed the real logic, is understandable. But it raises a whole host of issues, which have dogged the whole MFP debate ever since its inception. The most obvious is who is to pay, and how. South Australia, as a small state government with a population of little more than one million, is clearly not in a position to bear much of the burden itself, since its traditional economic base in primary production and import-substitution manufacturing has problems of its own. That suggests that the MFP will in effect become a kind of Japanese enclave in the middle of Adelaide, an extension of a process of Japanese investment in Australian urban infrastructure which has already grown by leaps and bounds since the early 1980s. There are of course benefits for South Australia in the form of privileged access to Japanese capital and technology, and also to the Japanese art of management; the scheme could just prove to generate the synergies that both sides clearly desire.

There is another curious aspect. Like the Japanese home-grown variety, the MFP has a Utopian element: the achievement of a perfect city, without barriers of nation, race, and class, and full of joy and creativity. This harks back to very old beliefs: that Japan is a "family state" and even a divine state, harmonious and virtuous; and that Japan can create a harmonious, equal "Commonwealth" with its neighbors, a notion assiduously propagated by Japan during the 1930s, now being revived in the face of new free-trade blocs in North America and Europe.[60] Certainly, this aspect of the scheme can be understood on at least two levels: first, as a piece of high-level international diplomacy, designed to enhanced MITI's status within the Japanese governmental machine; and second, as an exercise for Japanese internal consumption, intended to placate rising Japanese popular concern about the relatively poor quality of everyday urban life.

Whatever the case, the MFP is certainly a very exotic project; and its exoticism has been enhanced by the fact that both the proponents of the scheme, and its opponents, have succeeded in hyping it almost out of all recognition:

> Great claims have been made for the historical significance of the proposed Australian Multifunctionpolis. To supporters, it is "one of the most forward looking projects currently on the world horizon," and "may well be the most important single development in Australia as we head into the 21st century"; to opponents, it represents a realisation of the nightmare predictions of Aldous Huxley in *Brave New World*. To the casual observer, however, the most interesting thing about the MFP is perhaps the fact that so insubstantial a project should have become the object of such inflated hopes and fears.[61]

In fact, as one Australian-Japanese critic, Yoshio Sugimoto, has described it, it is "a moving, amorphous phantom the exact shape of which still remains unclear."[62] Its very vagueness, which has persisted since its first gestation, is a testimony perhaps to the fact that there is an underlying agenda, perhaps to the fact that MITI is embarked on a plan too subtle for Western comprehension. Or, if one prefers, there is an extremely simple agenda: to promote a Japanese construction bonanza. Or perhaps, all of the above.

CONCLUSION: THE POLITICAL FOUNDATIONS OF TECHNOCITIES

In some senses, these two projects are very different. One was the result of Spanish-British academic collaboration; the other was generated by the highly-creative minds of MITI's bureaucrats. One was a very specific response, carefully crafted, to the development situation and development problems of a distinctly backward yet highly-dynamic region on the European periphery; the other was a generic project for Japanese-Australian technical cooperation, which in a sense could have been located almost anywhere in Australia, and in fact shifted location several times during the period of gestation. In consequence, one is very specifically based on the very special circumstances – economic, social, cultural, political – of a highly distinct city and region; the other is much more generalized in character, appearing almost as an all-purpose blueprint for Japanese techno-colonialism in the twenty-first century.

Yet, despite these obvious differences, there are also similarities. Both are based on the need for international cooperation. Both are postulated on the notion of complex cooperative investments between the public and private sectors, and between indigenous and outside capital. Both are huge in scope, involving a mixture of fundamental research in university or institutional environments, and applied research and development in the laboratories of large private corporations. And both assume that the products of the research and development process will become exports to other regions and other countries. In short, both assume a globalization of the process of innovation, akin to the globalization of production that has become such a marked feature of the world economy of the 1980s and 1990s.

In terms of the physical strategy, both assume the development of a very large tract of urban land, effectively in the middle of a medium-sized metropolitan area. Both the urban areas concerned, coincidentally, could be regarded as peripheral in terms of the main centers of economic gravity of their respective countries. But both possess certain advantages, in that they are not over-developed or over-congested; yet both offer well-developed regional infrastructure in the form of airports, highways, and, above all, telecommunications, as well as a quite highly-developed local university base with a strong emphasis on engineering. Both strategies work on the basis of establishing certain key elements on the site, but then assume a degree of flexibility in response to changing circumstance. Both, in other words, are examples of strategic planning in uncertainty – as, indeed, is every other example considered in this book.

They are very large and ambitious projects; some would say so large as to be almost foolhardy. That, we would argue, misses the central point. In order to build any prospect of establishing a genuine innovative milieu, competing with established centers of technological innovation, such projects almost have to be launched on a huge scale. And, at any rate in the early days, they will inevitably have a speculative element. That is why they needed to involve the full faith and support not only of regional or state governments, but also of national governments – something that failed in Seville, and is not guaranteed in Adelaide. Furthermore, even this may not be sufficient without the prospect of outside support either from another national government, or from supranational agencies, like the European Commission.

Even more so than the other projects considered in this book, if realised they would take time to fulfill their potential – and even to demonstrate that the potential is indeed there. So a verdict on them will be due perhaps in the year 2010, at the earliest. They could demonstrate potential results before then, in the form of physical developments on the ground; but the crucial judgment, as to whether they constitute milieux of innovation, may have to wait that long. And since most political and economic actors are not interested in such long-term perspectives, it is highly unlikely that the blueprints underlying these projects will ever be fulfilled. Thus, their lessons could lie more in their failure than in their success. Technodreams can become technocities only if governments and corporations have a vision of the future, want to follow it, and marshal enough political support to endure the speculative moves and political manoeuvers that will undoubtedly try to derail the project for the immediate benefit of shortsighted personal interests. As all major projects of innovation, the construction of technocities is necessarily embedded in the political battles of the city, of the region, and of the world.

9

DISTILLING THE LESSONS

Technopoles have captured the imagination of national policy makers and local boosters alike. They have been hailed as the panacea for countries, regions, and cities struck by the painful adjustments needed in an era of technological and economic transition. They have also been denounced as ideological myths by those who yearn, nostalgically, for the return of a fast-disappearing industrial era.[1] As usual, life is more surprising, and not reducible to such unilateral views of social change. Our studies across the world yield a more mixed assessment of the potential of technopoles as engines of development. They have shown their diversity, and therefore the diversity of the criteria by which they must be judged. Our inquiry has also revealed the critical elements that seem to operate in the development of different types of technopoles, and the processes through which these come together, so creating synergy and spurring technological innovation and self-sustaining economic growth.

To be sure, there is no general formula for a successful technologically-based development, particularly since – under such evocative image-building concepts as technopole, or technopolis, or technology park, or science city – in reality they prove extremely diverse. Yet, because the various types of technopoles do articulate the productive forces of the new technological era, we do need to look into the preconditions for their formation. For technopoles provide the critical indicators of the territorial organization that will surely underpin the industrial structure of the emerging twenty-first century informational economies. On the basis of our worldwide survey, we can now address some critical issues that shed light on the genesis, structure, and outcomes of the emerging innovative milieux, that occupy the core of this new industrial space.

THREE FACES OF THE TECHNOPOLE

The first important finding is that there is not one motive, not one single objective, for pursuing a technopole policy; there are three. They can be seen almost as three corners of a triangle. Though they relate, sometimes confusingly, these three are really quite distinct from each other. And they need to be carefully distinguished, because they have different implications for any strategy of technopole building.

The first is, quite simply, *reindustrialization*. We say *re*industralization, because in most cases the objective here is to create new jobs in new industries, to replace old jobs in old industries that are contracting. In some, still-developing, economies, these are basic agricultural jobs that are disappearing as a result of land reform and greater rural productivity. In older, more mature economies, they are jobs in older basic extractive and manufacturing industries, resulting from earlier epochs of industrialization: industries that range from coal mining, through iron and steel production, to shipbuilding and even automobile manufacture. Product-cycle and comparative-advantage theory suggests that as economies develop, they should get out of industries where others can effectively compete, into those where there is some kind of advantage. This invariably implies moving up the skill-learning curve into more sophisticated products made in more sophisticated ways. It is the essence of Joseph Schumpeter's celebrated concept of creative destruction. In his formulation, the brutal disciplines of the competitive market would guarantee that this happened. But the process can also be planned in advance, to ease the strains. Further, in some cases, if such a process is not deliberately sought, it will not happen. Thus, the less developed an area, the more it will have to rely on a strategy of innovation-led reindustrialization.

The second objective is *regional development*. As nations and regions develop, moving up the curve as the first objective commands them to do, so they are likely to develop increasing disparities between one geographical area and another. In particular, the newer industries are liable to develop in one core region, where they draw on agglomeration economies by being close to one another. The older industries in contrast will either be agricultural, in which case they will be widely spread everywhere except in this more urbanized core region; or they will be older manufacturing industries based on exploitation of localized resources, like coal or iron ore or tidal water, in which case they will be concentrated in a few key regions or even key cities. In either event, increasing differences are likely to emerge in statistical indices of development as between one region and another, whether these are point-of-time indices like regional per capita incomes or unemployment rates, or change-over-time indices like employment or income growth. Such differences may be seen as bad on simple grounds of equity; but they may also be viewed as harmful in terms of national efficiency, since the continued growth of the successful core region may be accompanied by what the economists call negative externalities, or social costs, like traffic congestion, inflated housing costs, and air pollution. So the objective of national reindustrialization will be supplemented, or modified, by a regional goal: to concentrate the process in those regions that appear to be most in need. Almost invariably, this will mean trying to ease the growing industries out of the core region or regions, and into the less developed ones.

This can be done, and often is done, by a bundle of incentives to locate elsewhere and, sometimes, disincentives to further location in the core region. Firms, especially larger and more established ones, will respond to these by locating part of their operations – usually, new operations, arising from their continued growth – as branch-plants in peripheral regions. Such imported high-technology development can revitalize the economy of declining regions and can also provide the basis to diffuse new technologies throughout the traditional industrial fabric. But the overall control of the operation will remain in the old location, where the headquarters office is invariably found. And, since the research and development function is invariably found close to the headquarters, the source of further technological advance is still locked into the core region.

For this reason, there is a third and rather crucial element in the construction of technopoles – even though not all technopoles can hope to achieve it, and perhaps none can achieve it very quickly. This third element is the creation of synergy. Synergy is a word that is much used in recent literature on innovation, but that is difficult to define. It can best be regarded as the generation of new and valuable information through human interaction. The most spectacular examples of synergy are breakthroughs into new and important technologies, as when Shockley, Bardeen and Brattain together invented the transistor at Bell Laboratories in December 1947. But synergy could be less dramatic than that, and might not necessarily involve technological innovation at all. The creation of the package tourist industry, which depended on a new but already-existing technology (the jet airplane) and was essentially an organizational innovation, was one such example that involved a number of people both in demand-side countries like Great Britain (who, needless to say, ultimately controlled the process) and supply-side countries like Spain.

SYNERGY AND THE INNOVATIVE MILIEU

Many examples of successful synergy involve a combination of innovations – in product, in the organization of production, and in the penetration of a new market. The development of the personal computer, between 1974 and 1981, is a classic example, involving scores of people and companies in the San Francisco Bay Area. For this reason, following the original example of the growth of Silicon Valley, synergy is very often seen in terms of networks connecting individuals in many different organizations – public and semi-public and private, non-profit and for-profit, large-scale and small-scale – within a system that encourages the free flow of information and, through this, the generation of innovation. Such a place is the archetype of the innovative milieu, which can be defined as a place where synergy operates effectively to generate constant innovation, on the basis of a social organization specific to the production complex located in that place.

The Silicon Valley model has been so dominant that it has become a kind of myth, suggesting that it alone is capable of producing real synergy: that it, and only it, is the true innovative milieu. But the evidence in our case studies suggests that this is a serious error, for several important reasons. The first is that the kind of place that Silicon Valley represents – a new, previously non-industrialized region, not even fully urbanized at the point of first development – is not at all the only kind of innovative milieu in history. As this book has been at pains to stress, by far the commonest location of innovative milieux, especially in earlier times, has been in the hearts of the great metropolitan cities: London, Paris, Berlin, New York, Tokyo, and latterly Los Angeles and Munich. Though occasionally one of these might collapse through the fortunes of war (Berlin) or the fatal rival attractions of paper capitalism (New York), and though some others seem to have lost some of their synergistic capabilities in recent decades (London), others have remained highly innovative (Tokyo, Paris, Los Angeles). And, though Los Angeles like Silicon Valley is a relative urban upstart, the others are old places. You do not have to be young to be innovative; almost the reverse.

Both new and old places are at least visibly networked. They well correspond to that classic passage of Marshall in the *Principles*, which repays close analysis:

> When an industry has chosen a locality for itself, it is likely to stay there long: so great are the advantages which people following the same skilled trade get from near neighbourhood to one another. The mysteries of the trade become no mysteries; but are as it were in the air, and children learn many of them unconsciously. Good work is rightly appreciated, inventions and improvements in machinery, in processes and the general organization of the business have their merits promptly discussed: if one man starts a new idea, it is taken up by others and combined with suggestions of their own; and thus it becomes the source of further new ideas. And presently subsidiary trades grow up in the neighbourhood, supplying it with implements and materials, organizing its traffic, and in many ways conducing to the economy of its material.[2]

Latter-day analysts, discovering the virtues of vertically-disintegrated production systems and flexible specialization, are apt to point to this passage as illustration of the nineteenth-century industrial district, which has been reinvented in the late-twentieth-century Southern Californian aerospace complex. And indeed, within Marshall's concept of agglomeration economies are found the germs of much later key neoclassical economic insights concerning the significance of transaction costs and the resulting merits of markets versus hierarchies. But Marshall is going beyond even that: escaping from the constraints of neoclassical equilibrium analysis, he is anticipating the dynamics of Schumpeter's concept of innovation.

And even beyond Schumpeter: for Marshall correctly saw, what others were to see much later, that the process of innovation consists not in a single heroic act by a Schumpeterian "new man," but in a continuous chain or *cascade* of innovation set in motion by that first act. Such a cascade is already found in Lancashire in the half-century between Crompton's mule of 1779 and its final outcome, Roberts's self-acting mule of 1828. It is found in Silicon Valley in the 18 years between Shockley's arrival in 1955 and the perfection of the Intel 8080 integrated circuit in 1973, or in the even briefer period of gestation of the personal computer between 1974 and 1981. It is the essence of the synergistic process, and – as both these examples show – it is most likely to happen in dense urban networks. But – as again Lancashire and Silicon Valley demonstrate – not always established ones: they may be ones in course of creation, at the edge of the then-industrial world.

The notion of innovation chains suggests rather strongly that there need to be continuing close links between R&D and production. The product is tested and refined; a rival producer thinks of an improvement and brings it to market; the original producer reacts in turn, with yet another refinement: this is the essence of the Marshallian process. But it is somewhat contradicted nowadays, by the increasing evidence of division of labor by process: if production can be decentralized to distant regions and even distant countries, how is this reciprocal relationship assured? The answer in the Japanese case seems to be this: a special kind of factory, which Mitsubishi call their brain factory, is established close to the R&D center; this allows constant testing both of product and of process improvements. But it does mean that these two have to be kept in close juxtaposition, presumably in the innovative core region. And if then different products are decentralized to different regions, the question is whether all these research–production relationships be similarly separated and hived off: a daunting prospect, even for a large and diverse corporation.

THE CORPORATE INNOVATIVE MILIEU

Yet, looked at more closely, the notions of synergy and of the innovative milieu are more complex even than that. As Schumpeter pointed out, in the twentieth century the nature of innovation has changed. It need not, and now commonly does not, consist in an individual inventor making a major advance and then interacting with other inventors in a network; since at least 1900, with the growth of big industrial laboratories in major corporations, it has become routinized and bureaucratized. A company like IBM, headquartered at Endicott in upper New York State, is clearly and outstandingly innovative, yet it has no need of an urban network to generate synergy; it does so inside its organization. A similar pattern characterizes the operations of AT&T's Bell Laboratories in Murray Hill, New Jersey, or Motorola in Phoenix, and of Texas Instruments in Dallas. In terms of innovation, as the economists say, it internalizes its externalities. This model, first developed by American high-technology companies like Bell and General Electric, has been borrowed and further refined by their major Japanese competitors; Japan's outstanding record of industrial innovation, in the post-World War Two period, has been based not on a Silicon Valley model of synergy but on organized research within the laboratories of industrial giants like NTT, NEC, Mitsubishi, Toshiba, and Hitachi.

But with an important difference: whereas the American laboratories early fled the metropolitan milieu – General Electric for Schenectady, as early as 1900; Bell Labs for Murray Hill, New Jersey in 1941 – their Japanese equivalents have from the start remained firmly locked into metropolitan Tokyo. And there is a reason: these companies, based as they are on hierarchical organization and lifetime employment, nevertheless do relate to each other at some higher level. Above all, they relate to government. As part of Japan Inc., their competition – real, and even ferocious, as it is – is orchestrated within rules that are set by implicit agreement and adjustment between them and MITI, the Ministry of International Trade and Industry. Thus MITI, in launching its program for new technologies in the 1980s, invested in basic research in national public institutes – but then, as it had done before, encouraged the big companies to take up the further development and to divide the market among them.

This model, which we will call *statist corporate* as distinct from the American *private corporate* model, helps to explain the vital difference in location. Japanese companies lock their core activities, including R&D, into Greater Tokyo, not only because of their close financial relationships with the *keiretsu* banks from which they derive their basic funding – a key factor in their ability to take a long-term view of R&D, untrammelled by short-run, bottom-line considerations – but also because of their need for constant and close contact with the MITI headquarters in Kasumigaseki, a mere ten-minute limousine ride from their headquarters. Latterly, some of them have taken a leaf from the American book: somewhat on the model of Bell Laboratories in 1941, they have decentralized their R&D to semi-rural secluded locations at the edge of the metropolitan area, especially to the hillier southern half of Kanagawa prefecture, where researchers enjoy an almost American ambience. But it is clear that this process has limits, set by the ability to communicate regularly with headquarters, and with key production units.

This total centralization of economic life, in (and from) Tokyo, makes it difficult to comprehend the development of a "Silicon Valley" in an emerging Japanese region. Yet this is precisely what MITI's technopolis policy is trying to achieve. The Japanese clearly yearn for the Silicon Valley archetype; even though their own very different model has been so successful, they feel that it lacks some critical innovative spark. And indeed, entrepreneurial history suggests that cities or societies dominated by large, vertically-integrated corporations tend to provide extremely poor nourishment for aspiring entrepreneurs; thus, if their basic industries die, there is no source of innovation through which replacement could come.

America too has its own version of statist corporatism, however. Oddly enough, it is Los Angeles. On the East Coast, firms were free to flee New York City because it was a pure financial and manufacturing center, without governmental links; and, until 1980, Washington DC did not attract industry. But in Southern California, from the mid-1950s onward, the Pentagon in effect created its own version of government–corporation synergy. The Air Force's Western Development Division in Los Angeles, linked through the systems house of TRW to a clutch of key contractors, and through them to a vast network of subcontractors, operated another version of the Japanese model. But with a critical difference: governed by the special rules of defense contracting, which specify only the best at any price, the synergies of the whole system were systematically driven to diverge further and further from market norms. Becoming uniquely effective at achieving the stated objectives of the military procurement agencies, they progressively disabled themselves from competition in the ordinary, outside commercial world.

This same model seems to have been true for defense contracting in London's Western Crescent, where it has been a major factor in Britain's relative failure in the commercial high-tech marketplace. It appears also to have been a dominant model for much of the contracting in the Cité Scientifique Ile-de-France Sud, where it produced a uniquely French variant of the statist-corporate innovative milieu: extending the principles of military contracting into the civil sector, therefore mission-oriented rather than profit-oriented, in consequence capable of state-of-the-art technological triumphs in fields as varied as high-speed trains, supersonic airplanes, and telecommunications, but with conspicuously little success – at least so far – in diffusing advanced technologies into the consumer-durable industries. The difference between the Parisian and the Tokyo milieux, in some respects so apparently similar, will bear deep reflection for the serious reader of this book.

THE STATE AND INNOVATION

These different examples do however help to make a more general point: in the late twentieth century, the State constantly intervenes in technological development, in various ways and with varying degrees of success. The models range from the command economy of the former Soviet Union, as in the case of Akademgorodok, through to the many and varied cases in which a capitalist state enters into relationships – sometimes procurer–client, sometimes strategic coordinator, sometimes a mixture of these two – with private corporations.

These models are particularly evident in the case of the so-called developmental state, which is so characteristic – albeit in subtly different manifestations – of Japan and of the four East Asian tigers (Singapore, Hong Kong, Taiwan, Republic of Korea). But in different measure they are also typical of many advanced countries, where national governments back high-technology projects for reasons either of national defense or of national prestige. Even in that apparent archetype of the free entrepreneurial model, Silicon Valley, it is evident that defense contracting in the early years was crucial to the rapid technological development and hence the crucial price/performance ratio of the semiconductor industry; doubtless the industry would have grown anyway, but without the artificial pressure from defense contracting – which gave it a head-start, decisive for competitive success once electronics technology was diffused worldwide – it is unlikely that it would have grown nearly as fast.

Indeed, it is not too much an exaggeration to say that these State–private sector relationships are now characteristic of the mature capitalist State; the only significant distinction is in the precise methods employed, and in their apparent efficacy. Evidently, success depends on the ability of the State to mediate its presence, through the companies, in the commercial marketplace. Japan, Inc. is of course the outstanding success case. But when government attempts total intervention by itself, without the aid of the companies, even Japan can make a serious mistake – as the case of Tsukuba shows; here, MITI seems to have learned its lesson in the technopolis program.

What the Government can contribute is its ability to encourage research and development that could not be justified in a normal commercial balance sheet, because it is too large-scale or too high-risk, or both. Japan, where this attitude is backed within the private sector by the ability of the *keiretsu* banks to take the long-term view, is again the classic case. Consider here a technology like magnetic levitation, which the Japanese have backed for decades in the belief that it would eventually create a major transportation industry, while other nations like the United States and Great Britain have walked away.

The discussion up to now has treated the "State" as if it were a monolithic entity. Of course, it is not. Within the State are found many different levels and segments, which may behave independently and may even come into conflict. Even central or federal government departments have their own agendas. Local governments have theirs, which may clash with each other and with those of the center. Universities, whether public or private, enjoy a high degree of autonomy. Cambridge University made an effective alliance with local government in the 1970s to develop high-technology research and development with industrial applications, even though central government policy had long been to discourage industrial development there. Japan's MITI was ingenious when, in the 1980s, it encouraged local prefectures – themselves classic examples of quasi-autonomous bodies – to compete with each other in its technopolis program.

Generally, quasi-autonomous bodies like universities appear to be better generators of synergy than non-autonomous government institutes, especially those that are defense-related. Neither the Government Research Establishments in London's Western Crescent, nor their equivalents in the Cité Scientifique Ile-de-France Sud, seem to have been very effective in sparking such synergistic relationships with private industry. In Tsukuba, the story was the same; here, as in France, the tight hierarchies in both the public institutes and the private corporations may have been to blame. And Korea's Taedock appears to be repeating Tsukuba's mistakes, 20 years later. Akademgorodok was essentially a complex of research institutes that failed to develop industrial links: their financing came from the Academy of Sciences, and they never linked with the Soviet industrial enterprises which depended on their respective industrial ministries.

UNIVERSITIES AS TECHNOLOGY GENERATORS

The role of universities seems to have been more critical in helping to develop technopoles, and particularly those technopoles that can be characterized as innovative milieux. But our studies also show that it takes a very special kind of university, and a very specific set of linkages to industrial and commercial development, for a university to be able to play the role it often claims to play in the information-based economy.

Universities have played a fundamental role in the development of some of the most innovative technological milieux, such as Stanford at the origin of Silicon Valley, Cambridge University or MIT starting the spin-off process in their areas of influence, or the catalytic function of the Ecole Nationale des Mines in the birth of Sophia-Antipolis. Yet other major universities, like the University of Oxford, the University of Chicago and the University of Tokyo, have never generated major technological centers; while others, despite their location in the heart of a metropolitan innovative milieu, have developed few industrial linkages: thus the Faculté des Sciences d'Orsay in Paris-Sud, or the University of Moscow, which is totally removed from the Soviet Silicon Valley, Szelenagrad, located a mere 15 miles (25 kilometers) from Moscow.

Universities in fact may play three different roles in the development of a technopole – though occasionally, in some privileged locations, the same university may involve itself simultaneously in all three (as was the case for Stanford University).

The first and most important role is to generate new knowledge, both basic and applied. In this sense, research-oriented universities are to the informational economy what coal mines were to the industrial economy. For various reasons, universities are better suited to this role than either private or public research centers. Private research centers cater to their parent corporations, either systemically or through consulting contracts. Public research centers are dependent on government rules, and are not open to competition, thus having little incentive to diffuse their research findings or to take up problems raised at the industrial level. In addition, very often they depend on military or government bureaucracies that require confidentiality in their research.

In contrast, research universities tend to use young researchers, many of them graduate students who quit the university after graduation, thus spreading the knowledge they have acquired. Such universities operate a diversified system of incentives and rewards that drives their faculty, even after achieving tenure, to continued scientific innovation. Of course, all these arguments do not apply to universities that are pure teaching factories, or those where a bureaucratic structure disassociates the reward system from scientific productivity. Such universities will be extremely unlikely to act as the generators of advanced technological milieux.

The second function performed by universities is the training, in both requisite quantity and requisite quality, of the labor force of scientists, engineers, and technicians, which will provide the key ingredient for the growth of technologically-advanced industrial centers. Firms can of course recruit their personnel nationally or internationally, but this is only easy if they are already located in an advanced urban-industrial area. For all start-up technological centers, the ability to build a local labor market of good-quality engineers and scientists is critical. A university may perform this function even if it does not serve a major research role; though it is much more likely to be successful, especially in training higher-level scientists, if it does.

Last but not least, universities may assume a direct entrepreneurial role, supporting the process of spin-off of their research into a network of industrial firms and business ventures. The clearest cases are those in which the university itself sets up an industrial park, as at Stanford or Cambridge. But it is also possible for universities to encourage and allow their faculty members to set up shop by themselves, either leaving the university to start a company, or to work part-time in both worlds. The lack of this capacity can prove a very important drawback. A telling case is the major difference in attitude towards business-oriented professional activity between MIT and Harvard, which actually led several Harvard faculty members to leave the university in order to develop the industrial applications of their research. Similarly, the rigid compartmentalization of Japanese universities, whose professors are forbidden to develop any interest in outside companies, may prove to be an insuperable hurdle for the next stage of innovation in Japan, when leading-edge Japanese companies will be unable any longer to depend on downstream innovation based on imported research. The general rule is that the more purely academic the university, the less likely its contribution to technopole development.

But there is another rule: universities can only play their innovative role if they remain fundamentally autonomous institutions, setting up their own research agendas, and establishing their own criteria for scientific quality and career promotion. "In-house" universities, or research programs entirely dependent upon an external funding source, are critically vulnerable to special-interest pressures, and will in the long term undermine their own research and training quality. Thus, autonomous research universities, vocational universities, and entrepreneurial universities, based on scholarly quality and academic independence, yet linked to the industrial world by a series of formal ties and informal networks, are fundamental sources both of new information and of the human capacity to handle it; they provide both the raw material and the labor force that technopoles need.

FINANCE, INSTITUTIONS AND THE INNOVATIVE MILIEU

In Tokyo, the headquarters of the great innovative corporations do not relate merely to the State; they are also closely integrated with their parent banks. That of course is a feature unique to Japanese industrial organization, which goes back to the Meiji restoration of 1868 and was broken only for that short period after World War Two, when the American occupying power tried unsuccessfully to break the links. But it does point to the fact that innovative industries must be nourished by capital. Established major corporations, which maintain their innovative potential internally, can achieve this by whatever mechanism is appropriate, whether bank funding or through equity. But new enterprises, without preexisting reputations, must find ways of generating funds to keep them alive.

Historically, this has been done in highly informal ways, usually by recruiting local capitalists to act as "sleeping partners"; this model was equally true of late eighteenth-century Manchester, early twentieth-century Detroit and mid-twentieth-century Palo Alto. For this reason, though all these three were classic examples of new industrial places, they had some preexisting economic substance; to misquote Gertrude Stein, there was a There There, in the form of capital generated in some previous cycle of accumulation. Young Lancashire innovators in the eighteenth-century cotton industry were financed not by banks but by the existing "manufacturers," who were essentially merchants operating a form of domestic production. A young Henry Ford, in the Detroit of 1900, readily found backers from the city's plutocracy, who had already made fortunes in mining, logging or manufacturing. Hewlett and Packard started with loans from their professor, himself the son of a professor and a person of some substance.

But, for the extended innovative chains that began to arise after World War Two, involving protracted research and development, such informal sources of capital supply could hardly be sufficient. Hence the rapid growth of the venture capital industry, whose origins are peculiarly associated with the growth of Silicon Valley. The evidence however is that the industry was not there at the start; it was actually attracted to San Francisco by the evidence that something important was already happening a few miles away.

Recent empirical work in fact suggests that in the United States, there is indeed a strong bias in venture capital investments towards the Northeast and Pacific, specifically to California and Massachusetts, and then to major urban concentrations of high-technology business. Places like New York City and Chicago, which are major financial centers, are nevertheless weak venture capital centers. In other words, venture capital goes where high-technology industry already is. Yet much of the increase in venture funds has come from insurance and pension funds, which tend to be managed from these older financial centers. It seems that the money flows from these places to intermediaries close to the source of the action. So success breeds success; there is concentration, not filtering down. America's Mid-West is under-supplied with venture capital, though in large measure this reflects the preponderance of California and Massachusetts.

THE SOCIAL ORGANIZATION OF TECHNOPOLES

Our studies show that technological innovation does not result simply from the addition of the necessary factors of production of high-technology industries. The critical synergistic effects depend also on specific forms of social organization and institutional support. Social networks, allowing for the informal exchange of technological information and for the interpersonal support of an entrepreneurial culture, prove to be essential ingredients for the formation of a self-sustaining innovative milieu. These milieux form quite spontaneously in the course of time, although they may be facilitated by spatial arrangements or social institutions such as private clubs and leisure activities.

However, the how-to-do-it school of high-tech literature has mistakenly associated the image of freewheeling exchange among innovative engineers with the Silicon Valley mythology. This, we think, ignores the special quality of the California culture, and thus makes a functional rule out of a specific pattern of cultural behaviour. We do not want to underplay the crucial importance of mechanisms for creative interaction among innovators, including the informal exchange of strategically important information. Yet

California is not the only innovative land on earth, nor is its particular brand of innovative milieu the universal model for adoption.

The Greater Tokyo area has proven, in the long run, as innovative as Silicon Valley, and Munich or Paris may do so in the future if they break loose from top-heavy corporate bureaucracies. Social networks are also important for the ignition of the Japanese genius for innovation, and they may even take place in bars, although under a different set of organizational and cultural arrangements. Social observers have noticed the "Thursday Night" tradition of Japanese corporate executives and engineers, when they are socially permitted to go home late and even to get reasonably drunk, often on a corporate account. We do have anecdotal evidence indicating that informal exchange of information and ideas, along with content less relevant for our research purpose, does take place in the Tokyo bars, just as much as in the legendary Mountain View bars and restaurants of Silicon Valley.

Yet there is a fundamental organizational difference, rooted in another form of social structure: in Tokyo, informal groups are not entirely informal, since they always belong to the same company, albeit sometimes to different departments of the same company. In addition, more formal but equally productive interaction takes place within Japanese companies, with engineers participating actively in innovation and quality control groups that cut across the usual lines of command of the company. Thus, informal and semi-formal networks place a crucial role in the generation of new, valuable information, but not necessarily along the lines of the individualistic Silicon Valley model.

The research literature, which is too easily prone to a romantic approach to the generation of milieux of innovation, tends to ignore another critically important form of interaction: the constant interaction between large companies and their subcontractors, both in the American and Japanese models. It is in this hands-on production relationship that the most critical learning process takes place, with commanding companies forced to surrender some of their advanced technology if they want their contractors to achieve products up to their own technological standards.

Thus, social networks are indeed essential elements in the generation of technological innovation, and they are the backbone of the social organization of any innovative locality. True, their genesis, structure, and composition will vary widely, according to the specificity of the local culture and the constraints of the specific institutional environment. Yet it remains the case that without an innovative local society, supported by adequate professional organizations and public institutions, there will be no innovative milieu. And without an innovative milieu the development of high-technology industries will contribute to regional development only within the heavy constraints set by the business cycles of industries that are likely to be highly volatile. There will be no possibility of truly indigenous growth, and thus no escape from the state of dependency on another region, another region's companies, and another region's innovative individuals.

SOME IMPLICATIONS FOR REGIONAL DEVELOPMENT

These relationships have some possible implications for the regional element in technopole history. For they may help explain why the large metropolis tends to be a dominant innovative milieu. An urban university in a large city that is not just a campus city, a city where there is feedback from industry, offers better prospects for synergy than an ivy-walled college in a remote small town. This suggests that, in order to escape from the circle, an element of deliberate will, of breaking the mold, may be required: new synergies will have to be forged, in places that do not exhibit them to the same degree. And that in turn implies that it may be easier to do so by means of short-distance decentralization that does not upset all the old relationships. This may be the secret behind the fact that Hsinchu, 40 miles (70 kilometers) from Taipei, was successful while Taedock, 75 miles (120 kilometers) from Seoul, was not – at least, so far. Cambridge, which is 50 miles (80 kilometers) from London, was an undoubted success – but only, it might be argued, when London's growth had rippled out that far. And Tsukuba, which was also only 30 miles (50 kilometers) from the metropolitan core, remains an anomaly. Perhaps timing was to blame; simply, Tsukuba came too soon.

But we need to relate this discussion to a foregoing one. If innovation – whether through networks, or through bureaucratic hierarchies – essentially consists in a continuous chain or cascade of cumulative improvement, then where does the origin of the chain lie? We know that it can be geographically displaced, as occurred when the transistor was invented at Murray Hill, in New Jersey, but the cascade effect took place first in Palo Alto and then in Tokyo-Kanagawa. Further, although manufacturing is progressively displaced to lower-cost locations either at home or offshore (Oita, Kumamoto; Singapore, Malaysia), it seems that the basic R&D remains in the (displaced) core locations. This then raises the basic question: how important is the distinction between *basic* and *downstream* innovation? Some Japanese observers, we have seen, are concerned that their country has so far been able to manage only the latter, not the former. If so, it has done very well out of downstream innovation; but, the next time around, this may not be sufficient. Abundant evidence – already presented in this book – suggests that a crucial element of MITI strategy is now to secure the beachheads for basic innovation: the National Research Projects, the technopolis program and the Multifunction Polis are simply alternative routes to this end.

This topic is indirectly related, of course, to the debate on long waves in the capitalist economy, which, long moribund, took on a new lease of life during the 1980s. Clearly, a believer in the theory should be extremely concerned about the locus of the next long-wave process – particularly as it suggests that we are now near the end of the fourth Kondratieff cycle and close to the start of the fifth. Some in Japan, for instance in the National Land Agency, do believe that the next long wave will come around 2000 and will be based on electronics, biotechnology, new materials, and new energy. Naturally, their plan is to ensure that the start of this next cascading process will be in Japan.

But it is not necessary to be a born-again Kondratieff believer in order to be concerned about the locus of innovation chains. Even if one believes that the innovative process is more or less random in historic time, with no periodicity at all, it is still important to ask where the next innovative process may begin. At present – despite European breakthroughs in some important scientific fields, such as nuclear fusion – the answer appears open as between the Pacific coast of the United States and the Pacific coast of Japan. It will take a major initiative to break this particular mold, but major efforts are taking place to do so: within Japan itself, in Japan in association with Australia, and within Europe. There could yet be surprises. And, if they happen, they would result from planned action.

THE IMPORTANCE OF TIME

The case studies strongly underline a point that should be intuitively obvious: technopoles are not built in a day. At very least, something like the Cambridge phenomenon seems to have taken about a decade – perhaps longer – to yield significant results. At the opposite extreme, something as ambitious as the Japanese technopolis program, which depends on the creation of new networks in places that never had them, may take 20 to 30 years before the full effects can be gauged. We would estimate these as a rough approximation to minimum and maximum time horizons. The majority of schemes will fall between them, probably toward the upper range: say, 15 to 25 years for the full impacts to become evident.

There is an interesting implication: that an over-hasty conclusion would be quite wrong. This of course is common to many kinds of more conventional investments, like roads or airports, which may appear as white elephants in their early years, but which afterwards abundantly justify themselves. But here it is of the essence, since the necessary linkages may be almost organic in character: they will need to be grown, and to achieve this could take literally a human generation. That has to be very much borne in mind while reading the case studies; because many of the programs and projects are relatively quite young, a definitive verdict could be premature. Further, in general, the verdict can only get better over time; networks and linkages, once established, are unlikely to wither.

A WINNING FORMULA?

As the stories in this book have amply shown, technopoles are highly diverse. In the best, most productive cases they combine the following: an innovative milieu; the capacity to reindustrialize on the basis of advanced, competitive firms; and the ability to decentralize from traditional core locations into new, more adaptive, more dynamic start-up regions and localities. Yet, in other instances, technopoles are simply instruments of renewed economic growth or of regional development, on the basis of innovation and technological diffusion derived from other centers, old and new, where such innovation takes place.

But all technopoles, in order to deserve that title at all, must articulate certain key features: some form of generation of – or access to – new, valuable technological information; a highly skilled labor force; and (a production factor that cannot be taken for granted) capital ready to take the risk of investing in innovation. And the organizational combination of such specific sources of capital, labor, and raw material can hardly happen spontaneously, following the logic of the market, particularly in areas that are just starting up as industrializing centers. Some form of institutional entrepreneurship, either government, non-profit, or private, must intervene in the process. Only this can create the starting conditions that can light the creative fire of self-sustaining synergy.

It is somewhere in this twilight zone – of facilitating technological innovation and business entrepreneurialism, without suffocating inter-firm competition and human imagination – that the magic formula lies: the leadership of a new technological era. As Spanish explorers sought untold treasure, regions and cities around the world now seek a similar goal, now redefined: the twenty-first-century Eldorado. Perhaps, like their sixteenth-century equivalents, they will find that the ultimate goal, the innovative milieu, will never materialize. But perhaps, like them, they will discover a continent in the process. Or, at least, a modest piece of estate that will generate new wealth from information.

10

BUILDING TECHNOPOLES

How, then, to build the new Eldorado? How exactly does a nation, a region, a city, seek to determine its own technological-industrial future? How, by deliberate action and with foresight, does it create places that will prove especially hospitable to leading-edge industries based on the newest technologies, the kinds that will launch it into a successful economic future?

First, though, an even deeper and more disturbing question: is such an audacious enterprise at all possible? For our study has recorded many partial failures and many still-open verdicts. It has also shown that the uniquely precious and elusive condition, which we call the innovative milieu, is still to an extraordinary degree locked in the great metropolitan cities, where it has been found since the start of recorded economic history. Is it possible, then, that the entire technopole project is a chimera, dedicated to reversing deep currents of history, that prove in practice irreversible?

Such a deeply pessimistic conclusion can be rejected straightway. The birth of Silicon Valley in a previously unindustrialized area, that apparently had little in its favor, is one powerful piece of refutation; the revival of Greater Boston, an archetypal old industrial area whose decline appeared terminal, is another. And, against the partial failures, we can set the partial successes: new industries, new firms, new jobs created in unlikely places, some of them singularly unpropitious.

But the failures and the half-successes should be kept in mind. No one starting to build a technopole should be blind to the historical realities. Technopole construction is not an easy task, and the way forward is not always clear. The remainder of this chapter tries to provide a map and compass.

SETTING THE GOALS

The first essential, surely, is to determine the basic objective. As Chapter 9 started by showing, there are three possible aims of a technopole policy: to develop new industries as a national policy; to regenerate a declining or stagnant region; and to develop a milieu of innovation. These are by no means the same; indeed, in a particular case, they could well be contradictory.

A first example: a national industrial policy, in a developing country, might involve importing advanced technology into the national capital region, where the best-qualified labor force and the most advanced infrastructure are concentrated; that would do nothing for the other regions, and would not serve the creation of an innovative milieu. A second: a regional policy in any country, but especially a still-developing one, might hinder rapid national industrialization, by diverting resources away from the region or regions where they could be most productively used, into regions where their marginal return was far lower. And a third: an attempt to create a milieu of innovation might entail a slow process of building a major university or research institute, which would do little to help national industrialization or regional development in the short, or even the medium, term.

The choice of priority is therefore critical. It does not have to be all-or-nothing. But it does involve a trade-off between national and regional priorities, and between short-term and long-term objectives. And the choice will have profound implications for every aspect of technopole building: for the relationship to overall economic development strategy; for the investment in associated infrastructure, both "hard" and "soft"; and above all for location policy.

OVERALL DEVELOPMENT STRATEGIES

In general, economic theory – both of the neoclassical and neo-Marxist variety – suggests that an early-developing country will need to concentrate on importing existing technologies, by encouraging inward investment via transnational companies. At a modest level of development, it will be involved in competition with other countries for this pool of investment, and it must make itself noticeably attractive to it. Achieving a higher developmental level, when industrialization is already general, it should aim to develop sufficient technological capacity so as to improve on this imported technology, by a combination of downstream product innovation and process innovation; fine-tuning, in other words. Countries at this level of development need to apply their new-found technological ability to improve and rejuvenate existing industries that may suffer from existing or potential failure to compete. These may include agriculture and forestry, craft production, older basic manufactures, and services such as tourism.

Developed nations need a different strategy. They must continue to advance upstream, away from importation and improvement of other countries' technologies, and into development of their own. Basic scientific research and its technological application now becomes much more important, and encouraging it must become a primary aim of government policy. Particularly vital, here, is the identification of key technologies which will become the platforms for whole new industrial clusters.

Successful countries, the ones that have become examples in the textbooks of development, have of course gone through these levels, or are still going through them. Germany went all the way from level one to level three during the nineteenth century, Japan during the twentieth. Singapore, Taiwan and the Republic of Korea seem to be somewhere between levels two and three; Malaysia and Thailand are at level one, and are poised to move to level two. The illustrations demonstrate one important point about these levels: they are in no sense deterministic stages of development. A nation might get stuck at one level, or perhaps even regress. Conversely, the newly industrializing countries (NICs) in East Asia, like Singapore and Korea, seem to have jumped from level one to level three in one generation.

Regions like nations occupy different developmental levels. And lagging regions may require policies appropriate to a lower level, while the national economy, dominated by its leading regions, may have reached a more advanced one. Thus it will be quite appropriate to try to "export" advanced-technology industries from leading to lagging regions, while the leading regions – and therefore the economy as a whole – are moving into the field of indigenous innovative capacity.

One lesson from the history of virtually all cases of successful development – and perhaps all cases of successful rapid development, like Germany in the nineteenth century or Japan and Korea in the twentieth – is that state intervention is crucial. But, as emphasized in Chapter 9, it has to be the right kind of state intervention. At every stage, we would characterize it thus: the State must provide the right environment and the right basis for the development of vigorous competitive private enterprise. This is sometimes described as "planning," but that can be a highly misleading word. As Friedrich von Hayek, that scourge of the command economy, wrote long ago in *The Road to Serfdom*:

> It is of the utmost importance to the argument of this book for the reader to keep in mind that the planning against which all our criticism is directed is solely the planning against competition – the planning that is to be substituted for competition . . . But as in current usage "planning" has become synonymous with the former kind of planning, it will sometimes be inevitable for the sake of brevity to refer to it simply as planning, even though this means leaving to our opponents a very good word meriting a better fate.[1]

Indeed. Japan and Singapore, in different but related ways, have given this "very good word" a "better fate." But it is important that in the process, they have produced something very different from conventional notions of planning. It is highly significant that in Japan, MITI's periodical ten-year plans are called "Visions." MITI's self-defined role is first to chart a long-term technological development future, second to jump-start it by public investment in basic research and development, and third to promote and coordinate the resulting commercial exploitation through vigorous competition within the private sector. Korea's Ministry of Science and Technology is following the same model.

This is a strategy perfectly attuned to Japan's present position as the world's number two nation in terms of technological development and industrial power. But so was MITI's earlier and very different strategy from 1950 to 1975, which was designed to promote Japan's process of historic catch-up with the then industrialized nations – which is exactly the place Korea is now. The objective, and therefore the methods, were different in each case.

Japan's experience since World War Two is the perfect case study of guided development through what Chalmers Johnson has called the "Developmental State." It is a model that deserves the closest study by any other nation embarking on the same path. Certainly Korea, which is pacing it up the development curve and seems to be doing so even more rapidly, has learned well from it – even though, in the process, it may have repeated some Japanese mistakes. But the first lesson is the need to identify the nation's position on this learning curve, and hence the objectives and strategy that are appropriate to moving from one level of development to the next – or, perhaps, even jumping over it.

Technopole building will play a different role at each of these levels. In less developed economies, science or technology parks will be constructed as magnets for inward investment. The stress will be on those factors that are likely to be most important to the investors: good buildings or building sites, in an attractive environmental setting; with excellent highway access and nearness to an international airport; with excellent international telecommunications facilities, perhaps through development of a teleport offering direct advanced communications to the wider world; with good quality housing for managers; and, above all, with easy access to a substantial pool of well-trained and motivated labor.

At the next level, the emphasis will shift. Though all the above factors will continue to be significant, they will be joined by others. In particular, the availability of a well-educated (as distinct from merely well-trained) labor force will become much more important. That implies widespread good-quality university education at least up to Bachelor's and even up to Master's level, as distinct from an emphasis on basic primary and secondary education. At this stage, too, the State may need deliberately to foster applied research in certain specific fields related to target sectors of the economy, for instance agriculture, craft production, or tourism. In order to do this, it will also need to develop basic university research. For that reason, a nation cannot begin too early to build up at least one major research university. Indeed, it should try to do so at early stages of development, in advance of apparent need. Good-quality research traditions cannot be built overnight; they may take decades to mature, and they must be ready as soon as a country seeks to develop its own innovative potential. Taiwan's Hsinchu Science Park is a good example of a park that incorporated top-quality university research from the start.

In the third stage, as suggested by the Japanese example (and indeed by the earlier example of nineteenth-century Germany), the role of the State becomes more complex and more subtle. It must seek to promote basic research and its applications in a number of key target areas. Though it may do so in different ways, the argument of Chapter 9 suggests that it would be wisest to do so through encouraging advanced-level university research in top-quality scientific and/or technological universities. This may entail either expansion and upgrading of existing institutions, or the creation of entirely new ones. In either case priority will need to be given to the chosen universities, which will be more generously funded and will enjoy higher prestige than the more run-of-the-mill institutions. Universities themselves may take the lead, as Stanford did in the 1950s and Cambridge did in the 1970s and 1980s. And if, like Stanford and Cambridge, they are established research universities, this strategy may be very successful. Even newer, less prestigious places, like several of the new British universities established in the 1960s, or Germany's Dortmund University established at the same time, have established very successful science parks. But there have also been a number of failures.

This depends in part on perhaps the hardest part of the entire enterprise: links will need to be forged between university researchers and industry. This, like other features of the strategy, will depend on what can only be called the academic and industrial cultures. In highly-individualistic societies – represented by the West in general and by the United States in particular – these can be established informally, through casual interaction, and through the encouragement by entrepreneurial universities of spin-off enterprises. In more formalized societies – represented by Japan in particular, and perhaps by other East Asian NICs in general – the linkages will have to be rather carefully built through intermediate or bridging institutions, of which the Japanese technopolis program offers many examples.

THE TRIGGERS TO INNOVATION

In later stages of development, the choice of target technologies and industries is crucial. Here, the world appears to have arrived at a rather crucial stage in the 1990s. For half a century or more, technological innovation in the advanced economies has been driven by military demands in an era of global conflict. First during World War Two, then on a much larger scale during the Cold War of the 1950s, government defense agencies formed a new kind of alliance with corporate producers of advanced technology to create what has variously been called the military-industrial complex or the Warfare State. This alliance in effect generated new industrial complexes in areas that had previously been lightly industrialized, like Silicon Valley and Los Angeles – the first examples of deliberate technopole building. Indeed, it profoundly changed the industrial map of the United States, shifting the balance away from the old industrial heartland and towards the nation's periphery: the rise of the gunbelt.[2] And similar effects were noticeable in Britain's M4 Corridor.

The question now, with the Cold War effectively over, is what new driver could take the place of the defense priority: what William James once called the moral equivalent of war. To be sure, there are huge global challenges, as daunting as the specter of intercontinental ballistic attack: the threat of global warming, the dire poverty of so large a part of the world's population, the ravages of AIDS and of hard drugs, the problems of environmental management, the huge projected growth in the old-age population, the threat of mass fanatic terrorism, aging infrastructure, mass homelessness. And the outlines at least of the next generation of formative technologies are already known to all: they include the convergence of computing, television and telecommunications to form multimedia; biotechnology; bio-electronics; energy generation and transmission; new materials; and high-speed ground transportation.

This does not exhaust the list, of course. And both nations and regions should try to identify niche technologies in which they are likely to prove especially proficient. This is particularly important at regional level, where industries can be developed to meet local needs and then made into export platforms. During the 1960s, as part of a deliberate attempt to diversify the economy of the old industrial Ruhr area of Germany, two new universities were established. During the 1970s and 1980s, they logically specialized in research on treatments to clean up degraded industrial environments. These spun off into local industries, encouraged by public policy in the form of a large technology park attached to one of them, Dortmund University. Now these firms are beginning to export their products and services to similarly-afflicted old industrial areas worldwide.

The same can even apply at national level. Two not so trivial examples: in the 1960s Japanese manufacturers developed fiber point pens, at first using traditional bamboo, for people wanting to practice the ancient Oriental art of calligraphy. From small beginnings in the Pentel sign pen, this became the basis of a revolution in the way the whole world writes. Then, in the 1970s, because Japanese offices found it difficult to adapt typewriters to the demands of the Japanese alphabet, electronics manufacturers began to improve facsimile technology, which was actually of nineteenth-century origin, by digitizing it. The product took off in a way they never expected, creating yet another global revolution in information exchange.

The examples are highly specific, but the moral is general and clear: products, which fill local needs, can often serve national and international markets too. The need is to develop a program of collaborative research, between the public and the private sectors, that seeks to resolve local problems or barriers, or serve specialized local submarkets; and then to look to the private sector to find wider markets.

IMPLICATIONS FOR LOCATION POLICY

The development of a strategy by stages, it must by now be evident, has important implications for both the form and the location of technopoles.

In the first stage of development, a nation will probably not have much choice as to location, because virtually all its eggs will be in one basket: its qualified labor force, its advanced infrastructural base in the form of airports, highways and telecommunications, its web of ancillary service industries, its best housing, and its richest social and cultural networks, are likely to be highly concentrated in its leading region, which is invariably the one around its national capital. This strongly suggests the development of science/technology parks within the national capital region.

However, this region may well already be starting to experience fairly massive negative economic externalities in the form of traffic congestion, long and arduous work journeys, housing costs, and air pollution, which may make it a less attractive location than it might potentially be. And these factors may well deter inward investors. In such circumstances the obvious trade-off is to encourage the development of one or more technology parks toward the metropolitan periphery, which should be in the most socially prestigious and environmentally favored sector, and should if possible be well-connected to national highways and to the leading national airport.

In stage two, the case of a somewhat more advanced but still-developing economy – the archetypal middle-income or newly-industrializing country – these regional disparities are likely to loom even larger, and the negative externalities in the leading region may now be perceived as serious enough to justify a positive regional policy. The point where this happens must be a matter of very delicate judgment. It also depends on a number of characteristics specific to the geography of the country.

Its outcome will be the planned development of technological parks, or even a full-blown technological city, outside the sphere of the leading metropolis. To be successful this policy will have to build, yet again, on an established base of infrastructure and skills. This will usually be found in one of the leading provincial cities, which will have grown to the point that it offers an adequate platform of transportation, communication, and specialized facilities including university education. Even though negative externalities may be lower here, they may be rapidly accumulating with continued growth; so, once again, a deliberate effort should be made to find a peripheral location within this metropolitan area, which is well-favored for further growth, and well-served by infrastructure. An existing institution, such as a local technological university, may be relocated here as it expands; or a completely new university, preferably one with a certain prestige, can be newly established. Here, a particular difficulty needs to be faced: prestigious universities tend to be old-established, and there is a great deal of inertia in the system; the rise of new universities, like Berkeley and Stanford and Cal Tech, to positions of prominence is a somewhat rare event. The ideal, therefore, would be to relocate a nationally prestigious university. But, as its alumni invariably wield a great deal of political clout, this may be hard to achieve.

In stage three, the locational considerations may again depend on the precise geography of the country. Traditional regional development theory, which suggests that regional disparities widen during development, also comfortably concludes that they will narrow again as the country attains economic maturity. At that point, infrastructural networks – highways, modern railroads, airports – will be widely diffused across the national economic space. High-quality services, both producer services (banking, ancillary services) and consumer services (ranging from schools and universities and hospitals to restaurants and boutique shopping) are no longer so concentrated in the national capital region, but are widely available in provincial cities. So it will be relatively much easier to promote provincial technopoles, but perhaps – as the pressures on the capital region weaken – less necessary.

However, there are now serious doubts about this rather complacent view of history, derived as it is from discredited versions of neoclassical equilibrium theory. Mounting empirical evidence, as well as theory, suggests that severe spatial disequilibria may persist long into economic maturity. As Chapter 7 showed, major metropolitan areas tend to retain their capacity as innovative milieux, even if the precise locus of the milieu may decentralize a short distance toward the metropolitan periphery. The innovative networks may actually widen and deepen within the expanding metropolitan region, rather than dispersing to provincial cities. Tokyo is the classic case; Los Angeles is another. In both, the resulting negative externalities may actually worsen, as certainly seems to have been the case during the 1980s.

It is no wonder therefore that Japan has chosen to pursue an aggressive policy of provincial technopole creation on a very large scale. Whether any other nation could or should afford such an attempt is a moot question. MITI itself did not envisage anything so large in its original vision of technopoles; the expansion was a result of political pressures. There are arguments both for and against.

The argument against is that such dispersion weakens the national effort that can be put into any one technopolis site, and so makes it less likely that any one of them will succeed. The argument for is that by encouraging local governments to compete by developing their own programs, a scatter-shot approach can actually increase the chances that at least one or two sites will prove really successful. We feel somewhat sceptical on that last point, and would be inclined to argue for greater concentration. But we do recognize the force of the contrary argument, and we are by no means certain; the proof of the pudding will be in the eating.

In any case, the technopolis program does offer clear implications for policy. It is openly based on a kind of two-level strategy. The first is to develop an advanced branch-plant economy in selected provincial regions, chosen on the basis of existing concentrations of basic facilities, by incentives to major corporations coupled with generous infrastructural investment. This is fairly traditional regional policy, albeit targeted on advanced-technology industries; it is not too different from what the British did in central Scotland, when they attracted British and American electronics plants in the 1950s. It can probably be successful in itself, at least in some areas, in generating new jobs; but it is essentially the kind of program we have suggested as appropriate for stage two of development.

The second however is novel: it is deliberately to build, on this traditional platform, a network of "soft" infrastructure intended to promote synergy and thus increase the capacity for indigenous regional innovation. In other words, it aims to do, on a huge scale, what has only been tried fitfully and partially elsewhere (at Sophia-Antipolis, for instance): to create a new milieu of innovation, where none existed previously. Such a strategy is entirely appropriate for a stage three economy. The combination varies somewhat in each case, but invariably involves expansion and upgrading of local universities; the development of new institutions, in the form of research centers, to provide the critical bridging function between academic research and industrial application, so lacking until now in Japan; and very generous physical networking, in the form of new structures for information exchange.

It is too early to say how successful this higher-level strategy may be. But, if there is an appropriate design for the deliberate creation of regional innovative milieux, it is probably bound to look something like the technopolis program. The only point of reservation must be that it is very culturally specific. The bridging institutions seem to be deliberately designed to overcome a well-recognized deficiency in the Japanese capacity to innovate. They may not be equally appropriate, at least in this particular form, to other cultures.

Latterly, however, Japan has gone beyond technopoles. The Multifunction Polis is in some ways the most extraordinary project described in this book, because it is so completely *sui generis* and so speculative. It appears to represent almost a fourth stage of development: the attempt, by an economically mature nation, to colonize innovation by going offshore to obtain it. Whereas in the nineteenth century developed nations sought colonies for their raw materials (and indeed, Japan did so down to the 1930s), now such a nation seeks a colony as a source of basic, upstream innovation.

The underlying logic is the same: just as a century ago developed nations lacked vital supplies of copper or rubber, so now some of them lack essential resources of information. As then they explored jungles and raised the flag over mines or tropical plantations, so now they seek to forge alliances with – and ultimately, perhaps, to attain control over – universities and research institutes in countries where the intellectual climate is favorable for the production of the new raw materials of the twenty-first-century economy.

If this is indeed the logic, there is no guarantee that it will work. And in any case, it is probably irrelevant to any other nation contemplating technopole building, because no other nation currently finds itself in Japan's advanced position on the learning curve, coupled with its particular set of concerns about its capacity for upstream innovation. Such a position might apply to the East Asian NICs about the year 2030, if they maintain their current growth path and momentum. By then, hopefully, the verdict on the MFP should be in.

THE SIGNIFICANCE OF THE TIME DIMENSION

"The verdict should be in." We have used that phrase more than once in concluding the case studies. And, in Chapter 9, we have emphasized the importance of the long-term perspective. It must be underlined again in the policy context. Building technopoles is not a policy that is likely to prove successful in the short haul. It will not yield dividends within the perspective of the short-run business cycle, or of the political election cycle. If it is to enjoy success, this will have to be measured over decades.

It does therefore need the kind of long-term commitment, not least on the part of government, that has not often been forthcoming – at least in Western cultures, with their stress on instant gratification in the economic and in the cultural sphere. It needs to be viewed in the same way as other elements of a development strategy, such as education and research, with which it is so intimately associated: as an extremely long-term investment, which will have to be nursed through early barren years. And this applies not merely to government, but also to the associated private investment. A way needs to be found, especially in Western economies, of encouraging the private sector to take the long-term perspective on research and development, which the Japanese corporations apparently find so natural.

This, we believe, has a great deal to do with the way that industry raises its finance. The Japanese *keiretsu* appear to be much more protected from the vicissitudes of the stock market, particularly its more speculative manifestations, than their Western counterparts. Leveraged buyouts, junk bonds and greenmail do not seem to be part of Japanese corporate life. Part of the explanation seems to be that more of the finance is raised internally, from the parent bank. Another part is that equity investors expect their return to come from long-term capital appreciation, not from short-term speculation. Both appear to be successful in circumventing the kind of predator capitalism that has been such a notorious feature of Western economies during the speculative mid-1980s. We doubt that a stage three technopole policy, the creation of a true innovative milieu, will be anywhere possible without a Japanese-style capacity for long-term investment.

TWELVE POINTERS TO POLICY

The implications for policy can be summed up in a number of aphorisms:

1 *Build a clear development strategy*. Different technopole policies are appropriate to different levels of development. The objective is to get from the stage of development currently reached, to at least the next rung of the developmental ladder – and, if at all possible, to jump more than one. At lower levels, relatively modest technology parks will be perfectly appropriate. Active regional policy enters at an intermediate level. Building true synergies will come at a relatively late, sophisticated stage. Consequently:

2 *Branch-plants are better than no plants*. All nations start by importing their technology. The only practicable way to do this is by encouraging inward investment and by training the labor force to serve this investment, simultaneously working to try to achieve a degree of indigenous competence.

3 *Synergy is crucial in the long run*. It is not the only objective of technopole-building, nor the only criterion of success; as just said, a technopole based on branch-plants, controlled from distant locations, will be a necessary first stage. But synergy, as the source of innovation, should be the long-run objective of any such program. And, therefore:

4 *Develop a long-term vision*. While there should be a due sense of realism about what is possible at any level, it will also be vital to develop a forward vision – particularly if the aim is accelerated development on the Korean or Singaporean model. For this reason, a strategy must early on seek to lay the foundations for advanced-level growth – such as the development of a major research university.

5 *Sources of innovation must be identified*. These need not be identical in every place. University or institute research may provide the basis in one country or region, not in another. Universities are usually good catalysts, but only when they are the right kinds of universities. National institutes, however prestigious, may do little if isolated from industry: a reciprocal relationship, research–production–research, is critical. The same limiting judgment may be true of private industrial laboratories, if they are totally isolated from networks. Therefore:

6 *Networks must be established early on*. There must be mechanisms and channels for information to flow. The free-exchange model of Silicon

Valley may not however be the only one. Some societies may achieve networking in a more formal, hierarchical way. It is therefore important first to understand how the national cultural system operates. Once that is done, it should be possible to determine how far it needs to change in order to achieve the desired degree of synergy. Simply providing the networks, in a physical sense, may not be sufficient; it may be necessary to take definite steps to open up the social networks and to break down barriers.

7 *Short-distance strategies may be easier.* In countries still at lower development levels, having a considerable concentration of institutions and networks in a central core region, it may be best to attempt local decentralization to the periphery of this region rather than developing distant regions, at least in early stages. Such a strategy could include movement of prestigious local universities and/or research institutes together with industry, at least the R&D component. But, if this is done, it is important that efforts are made to develop synergies from the start. Even short-distance moves can be very negative if the institutions do not talk to each other. It is particularly important that private laboratories be encouraged to move at about the same time as public research; otherwise there will be no spin-off by definition.

8 *Longer-distance strategies require selectivity.* If it is desired to build up synergistic capacities in other, more peripheral, regions, it will almost certainly be necessary to concentrate on one or two target areas that appear to offer the best prospects in terms of pre-existing facilities such as universities, industrial traditions, entrepreneurial capacities, and political leadership. A scatter-shot strategy, though more acceptable in political terms, may fail to build up momentum anywhere. But it must be recognized that these longer-distance impelled strategies are hazardous, as the example of Korea's Taedok illustrates.

9 *Major central inducements.* Defense spending played a key part in at least three of the case studies in this book (Silicon Valley, Los Angeles, London's Western Crescent) and quite probably others (Paris, Munich). Though that reflected the unique historical circumstances of the Cold War in the 1950s, it might be possible to replicate the experience with a similar national priority program. If, for instance, nations determined that the threat of global warming, or some other environmental hazard, were a priority of similar urgency, then the resulting R&D program could be diverted into a new region; and industrial development might well follow. The need, to repeat an earlier point, is to develop a long-term strategic vision, as MITI so successfully did in Japan. Though we now associate this with a leading-edge industrial strategy, in fact MITI also followed this policy at lower levels of development. It can in fact work at any level. Within this strategy, especially at regional level, the need is to:

10 *Identify new niches.* The Ruhr example shows how a region can develop specialized high-technology industries to meet local needs, and then exploit their export potential. However, a word of warning is due: such a strategy demands a particularly clear prior appraisal of the region's abilities to rise to such a challenge. There must be a basic physical and intellectual infrastructure – like the Ruhr's two new universities – already in place.

11 *Keep consistency.* The period necessary to achieve a successful technopole is a long one – certainly longer than that which most Western companies would allow before profit must emerge, and also longer than the political election cycle. Therefore protective mechanisms must be provided that will insulate the project, for a long period, against premature accusations of failure. Conversely, if part of the program is successful, every attempt should be made to ward off speculative commercial development pressures. A techno-park, for instance, should not be allowed to degenerate into a pure office park just because speculative profits are feasible. Protecting the integrity of the project has to be a first responsibility for state policy.

12 *The best may be the enemy of the good.* Countries and regions should not seek to judge all their efforts by the most rigorous, exclusive criteria. The axiom earlier suggested – branch-plants are better than no plants – is just one illustration of a more general rule: that even if a particular technopole policy is not entirely successful, it may have useful fall-out effects.

Finally, a word both of caution and of encouragement. There is no single path to the achievement of a successful technopole. Though there are general frameworks for policy, which are ignored at their peril, these are broad in nature: they allow, and even compel, a great deal of local variation depending on the specific national and regional circumstances. It is for each nation, each region, each city, to work out an appropriate strategy, with as much vision and as much imagination as it can conjure up.

NOTES

CHAPTER 1

1 There is however a problem, since in the French language the term has two different genders and meanings: *le technopôle*, derived from *pôle* (n.m., "pole"), and *la technopole*, derived from *métropole* (n.f., "metropolis").
2 Forester, 1987.
3 Ohmae, 1990.
4 Porat, 1977; Monk, 1989; Hepworth, 1989.
5 Solow, 1957.
6 Nelson, 1981; Denison, 1985.
7 Piore and Sabel, 1984.
8 Carnoy *et al.*, 1993.
9 Cohen and Zysman, 1987; Soete, 1991; Guerrieri, 1991.
10 Rosenberg, 1976.
11 Castells and Tyson, 1988.
12 Freeman, 1987; Saucier, 1987; Pavitt, 1988.
13 Dosi *et al.*, 1988.
14 Bijker, Hughes and Pinch, 1987.
15 Scott and Angel, 1987.
16 Goodman, 1979.
17 Massey, Quintas, and Wield, 1991.
18 Hall and Preston, 1988.
19 Schumpeter, 1942.
20 Castells, 1984; Andersson, 1985a; Arthur, 1985, 1986; Aydalot, 1986a, 1986b; Hall, 1990.
21 A reviewer has commented that we should properly begin with these innovative metropolitan cities, since they came prior to any of the policy experiments in the book. But, having pondered the matter, we think that the logic of our book demands that they be considered at this point and not before.

CHAPTER 2

1 Saxenian, 1993.
2 Rogers and Larsen, 1984.
3 Bernstein *et al.*, 1977.
4 Saxenian, 1990.
5 Saxenian, 1993.
6 Rogers and Larsen, 1984; Malone, 1985.

7 Saxenian, 1981, 1990; Scott and Angel, 1987.
8 Hall, 1995 forthcoming.
9 Rogers and Larsen, 1984; Malone, 1985.
10 Saxenian, 1990.
11 Mutlu, 1979.
12 Oakey, 1984.
13 Saxenian, 1990.
14 Saxenian, 1990.
15 Markusen *et al.*, 1991.
16 Saxenian, 1990.
17 Saxenian, 1990.
18 Saxenian, 1993.
19 Rogers and Larsen, 1984.
20 Saxenian, 1990.
21 Florida and Kenney, 1987.
22 Freiberger and Swaine, 1984; Hall, 1995 forthcoming.
23 Malone, 1985.
24 Saxenian, 1990; Borrus, 1989.
25 Kranzberg and Pursell, 1967.
26 Hall and Preston, 1988.
27 Rogers and Larsen, 1984; Keller, 1981.
28 Carey and Gathright, 1985.
29 Rogers and Larsen, 1984.
30 Azouaou and Magnaval, 1986.
31 Siegel and Markoff, 1985.
32 Saxenian 1981, 1985.
33 Kroll and Kimball, 1986.
34 Siegel and Markoff, 1985.
35 Saxenian, 1981.
36 Castells, 1989; Hall, 1995 forthcoming.

CHAPTER 3

1 This account relies heavily on the standard sources: Lampe, 1984; Loria, 1984; Saxenian, 1985; and Wildes and Lindgren, 1985.
2 On the first and second waves, see especially Dorfman, 1982, 1983; Fishman, 1981; Lampe, 1988; and Saxenian, 1985.

CHAPTER 4

1 Information and data contained in this section come directly from our field research, conducted in 1990 with the support of the Institute of Economics and Industrial Organization, Siberian Branch of the USSR Academy of Sciences, and with the research assistance of Ms Natalia Baranova. For general information on Siberia's regional structure and development policies see Schiffer (1989) and Wood and Joyce (1991).
2 We were seriously told that in the field of chemical engineering, for instance, one published scientific article per year would be sufficient.
3 Most of the information in this section comes from our own fieldwork in November 1988, in Seoul and Taedok. The inquiry was conducted with the collaboration of Professor Ju-Chul Kim, and the full support of the Korean Research Institute on Human Settlements.
4 Lee, 1988.

5 Lim, 1985.
6 Castells, 1992.
7 Richardson and Hwang, 1988.
8 Park, 1986.
9 Anon, 1990.
10 Park, 1986.
11 Anon, 1990.
12 Anon, 1990.
13 Tatsuno, 1986: 110–11.
14 Takahashi, 1981.
15 Tatsuno, 1986: 96–7.
16 Tatsuno, 1986: 97.
17 Kawashima and Taketoshi, 1987: 20–1.
18 Gakuen Toshi, 1985; Kawashima and Taketoshi, 1987.
19 Edgington, 1989.
20 Gakuen Toshi, 1985.
21 Onda, 1988: 58–62.
22 Gakuen Toshi, 1985.
23 Gakuen Toshi, 1985.
24 Gregory, 1985.
25 Gregory, 1985.
26 Bloom and Asano, 1981; Kawashima and Taketoshi, 1987.
27 Onda, 1988: 64–6.
28 Gregory, 1985.

CHAPTER 5

1 Sophia-Antipolis, 1989.
2 Sophia-Antipolis, 1989: 1–3.
3 Quéré, 1990.
4 Lafitte, 1985: 88.
5 We are indebted to Jean-Paul de Gaudemar for this point.
6 Quéré, 1990: 8.
7 Lafitte, 1985: 89.
8 Quéré, 1990: 8.
9 Sophia-Antipolis, 1989.
10 The category "Higher education and research" in Table 5.1 includes only those in research institutions; it excludes those in industrial laboratories.
11 Sophia-Antipolis, 1989.
12 Lafitte, 1985: 87–8.
13 Perrin, 1986a, 1986b.
14 Perrin, 1986b.
15 Perrin, 1986a, 1986b.
16 Lafitte, 1985.
17 Quéré, 1990.
18 Quéré, 1990: 20.
19 Perrin, 1988.
20 Perrin, 1988: 155.
21 Segal Quince Wicksteed, 1985.
22 Keeble, 1989: 156–8.
23 Keeble, 1989: 161–3.

24 Segal Quince Wicksteed, 1985: 15–17.
25 Segal Quince Wicksteed, 1985: 18.
26 Segal Quince Wicksteed, 1985.
27 Segal Quince Wicksteed, 1985.
28 Segal Quince Wicksteed, 1985; Keeble, 1989.
29 Segal Quince Wicksteed, 1985.
30 Keeble, 1989.
31 Saxenian, 1988.
32 Keeble, 1989.
33 Segal Quince Wicksteed, 1988.
34 Monck *et al.*, 1988, *passim*.
35 The trip takes about an hour on the highway, but it can be as long as two hours during the peak periods.

CHAPTER 6

1 Tatsuno, 1986: 121–3.
2 Broadbent, 1989: 235.
3 Murata, 1988: 49.
4 Sakamoto, 1988.
5 Douglass, 1988.
6 Johnson, 1982: 35–82, 195–7.
7 Johnson, 1982: 202–12, 273–4.
8 Tatsuno, 1986: 24–5.
9 Tatsuno, 1986: 27–8.
10 Okimoto, 1989: 67.
11 Tatsuno, 1986: 28–9, 40; Okimoto, 1989: 67–70.
12 Kawashima and Stöhr, 1988.
13 Industrial Structure Council, 1980; Tatsuno, 1986: 118–20.
14 Industrial Structure Council, 1980: 187–8.
15 Sase, 1982; Tatsuno, 1988: 1.
16 Sase, 1982; Tatsuno, 1986: 1, 47; cf. Furutate, 1982.
17 Douglass, 1988; Broadbent, 1989: 234, 237.
18 Broadbent, 1989: 238.
19 Murata, 1988: 41–2.
20 Imai, 1986: 153–5, 164.
21 Furutate, 1982.
22 Fujita, 1988: 568; Murata, 1988: 41.
23 Tatsuno, 1986: 128–32.
24 Fujita, 1988: 73, 568–9.
25 Noguchi, 1986.
26 Noguchi, 1986.
27 Noguchi, 1986.
28 Japan Economic Survey, 1984.
29 Tatsuno, 1986: 185–7.
30 Broadbent, 1989: 240–3.
31 Broadbent, 1989: 244–7.
32 Johnstone, 1988.
33 Fujita, 1988: 589–90.
34 Broadbent, 1989: 240.
35 Fujita, 1988: 574–7.
36 Tatsuno, 1986: 176–7; Johnstone, 1988.

37 Fujita, 1988: 589–90.
38 Tatsuno, 1986: 181–2.
39 Maruyama, 1985: 76.
40 Maruyama, 1985: 93.
41 Fujita, 1988: 578, 586–8; Tatsuno, 1988.
42 Maruyama, 1985: 76; Fujita, 1988: 578–80.
43 Maruyama, 1985: 81.
44 Johnstone, 1988.
45 Johnstone, 1988.
46 Maruyama, 1985: 81–3; Tatsuno, 1986: 183.
47 Nakano, 1988.
48 Murata, 1988: 50.
49 Sakamoto, 1988.
50 Sakamoto, 1988.
51 Sakamoto, 1988.
52 Broadbent, 1989; Douglass, 1988; Edgington, 1989; Glasmeier, 1988; Kawashima and Taketoshi, 1987; Kawashima and Stöhr, 1988; Masser, 1990; Toda, 1987; Tatsuno, 1986, 1988.
53 Edgington, 1989.
54 Edgington, 1989.
55 Nishioka, 1985.
56 Sakamoto, 1988.
57 Edgington, 1989.
58 Glasmeier, 1988.
59 Edgington, 1989; Tatsuno, 1990.
60 Furutate, 1982.
61 Glasmeier, 1988.
62 Glasmeier, 1988.
63 Edgington, 1989.
64 Roper-Dennis, 1984.
65 Glasmeier, 1988.
66 Hall, 1987: 337.
67 Douglass, 1988.
68 Edgington, 1989.
69 Yamasaki, 1990.
70 Edgington, 1989.

CHAPTER 7

1 Roe, 1916: 15–20, 33–46.
2 Martin, 1966: 35–6.
3 Martin, 1966: 38.
4 Smith, 1933: 105, 173.
5 Keeble, 1968; Hall *et al.*, 1987: 99.
6 Hall *et al.*, 1987: 114, 119–21.
7 Hall *et al.*, 1987.
8 Decoster, 1988.
9 Anon, 1989.
10 There is an uncanny similarity between the location of St Cyr in relation to the palace of Versailles, and that of Sandhurst–Camberley in relation to Windsor Castle.
11 Castells, 1967.

12 Decoster and Tabaries, 1986.
13 Decoster, 1988.
14 Itakura and Takeuchi, 1980: 47.
15 Murata and Takeuchi, 1987: 222–4, 228, 231, 237.
16 Ando, 1983: 3–4.
17 Kanagawa, 1988: 368–9.
18 Kanagawa, 1988: 369.
19 Murata and Takeuchi, 1987: 232–5.
20 Ando, 1983: 16–18.
21 Kanagawa Prefectural Government, 1985: 125–7.
22 Kanagawa Prefectural Government, 1985: 152–4.
23 Kanagawa Prefectural Government, 1985: 192.
24 Ando, 1983: 14.
25 Itakura and Takeuchi, 1980: 52; Ando, 1983: 16–18; Kanagawa Prefectural Government, 1985: 174–8, 192–5.
26 Itakura and Takeuchi, 1980: 51.
27 Itakura and Takeuchi, 1980: 51–2, 59.
28 Kanagawa Prefectural Government, 1985: 271–2.
29 Fujitsu Corporation, 1986; Nihon, 1972; Tokyo Shibaura Electric Corporation, 1977, *passim*.
30 Kanagawa Prefectural Government, 1985: 291–4.
31 Hasegawa, 1983: 262–3.
32 Kanagawa Prefectural Government, 1985: 291.
33 Ando, 1983: 14–15.
34 Hasegawa, 1983: 263.
35 Kanagawa Prefectural Government, 1985: 300–2.
36 Kanagawa Prefectural Government, 1985: 317.
37 Ando, 1983: 15–16.
38 Kobayashi, 1983: 152.
39 Takanashi and Hyodo, 1963: 59–61.
40 Takanashi and Hyodo, 1963: 69.
41 Kanagawa Prefectural Government, 1985: 321.
42 Kanagawa Prefectural Government, 1985: 321.
43 Takanashi and Hyodo, 1963: 106–7.
44 Ando, 1983: 16.
45 Miyakawa, 1980: 270.
46 Miyakawa, 1980: 271–2.
47 Kanagawa, 1988: 373.
48 Kanagawa, 1988: 371.
49 Itakura and Takeuchi, 1980: 59.
50 Itakura and Takeuchi, 1980: 48, 52–3.
51 Itakura and Takeuchi, 1980: 53–4.
52 Itakura and Takeuchi, 1980: 59.
53 Itakura and Takeuchi, 1980: 54–5.
54 Itakura and Takeuchi, 1980: 60.
55 Kobayashi, 1983: 160–1.
56 Kobayashi, 1983: 153.
57 Kobayashi, 1983: 153–4.
58 Kobayashi, 1983: 159–60.
59 Itakura and Takeuchi, 1980: 60–1.
60 Klingbeil, 1987a; 44–5.
61 Grotz, 1989: 267–8.

62 Klingbeil, 1987a: 47–8; Bavaria, 1988: n.p.
63 Klingbeil, 1987a: 48.
64 Bavaria, 1988: n.p.
65 Bavaria, 1985: 3–4.
66 Bavaria, 1987: 3–7.
67 Bavaria, 1988: n.p.
68 Klingbeil, 1987a: 46–7.
69 Klingbeil, 1987a: 45–6.
70 Klingbeil, 1987a: 46.
71 Klingbeil, 1987a: 52–4.
72 Klingbeil, 1987c: 129.
73 Klingbeil, 1987b: 74.
74 Klingbeil, 1987b: 71–3.
75 Klingbeil, 1987b: 74.
76 Klingbeil, 1987b: 74.
77 Klingbeil, 1987a: 48, 65.
78 Thürauf, 1975: 39.
79 Klingbeil, 1987b: 75.
80 Klingbeil, 1987b: 76.
81 Klingbeil, 1987b: 71–2.
82 Klingbeil, 1987b: 77.
83 Hall and Preston, 1988: chapters 8 and 13.
84 Thürauf, 1975: 43–4.
85 Siemens, 1957b: 271–2.
86 Siemens, 1957b: 274.
87 Klingbeil, 1987a: 48; Weiher and Goetzeler, 1981: 128.
88 Siemens, 1957b: 282; Weiher, 1987: 161.
89 Goetzeler and Schoen, 1986: 64, 72, 77.
90 Klingbeil, 1987a: 48.
91 Klingbeil, 1987a: 65.
92 Klingbeil, 1987a: 46.
93 Delius, 1977. The Delius account, ironically described as a *Festschrift*, was the subject of a prolonged legal action by Siemens from 1972 to 1975. As a result several statements in the book were ordered to be withdrawn, and so appeared in subsequent editions as "censored" by heavy black bars. The statement about the secret Yalta map was not however one of these.
94 Schreyer, 1969.
95 Klingbeil, 1987a: 56–7.
96 Klingbeil, 1987a: 65–6; Klingbeil, 1987b: 79.
97 Klingbeil, 1987a: 66.
98 Giese, 1987: 250–1.
99 Grotz, 1989: 268.
100 Bavaria, 1985: 5.
101 Giese, 1987: 250–1.
102 Bavaria, 1985: 6.
103 Scott and Mattingly, 1989.
104 Scott, 1989b.
105 Scott, 1989b.
106 Scott and Paul, 1990.
107 *Hell's Angels*.
108 Soja, 1986.
109 Markusen *et al.*, 1991: 100.

110 Hall, 1988.
111 Scott, 1986a, 1986b.
112 Hall and Preston, 1988.

CHAPTER 8

1 Castells and Hall, 1992.
2 Hamilton, 1991: 9.
3 Hamilton, 1991: 10.
4 Hamilton, 1991: 30.
5 Hamilton, 1991: 9, 30–1.
6 McCormack, 1991b: 50.
7 Inkster, 1991: 27; cf. McCormack, 1991b: 39.
8 Sasaki, 1991: 140–1.
9 Inkster, 1991: 27. Curiously, two sections of the report – on space industries and energy development – were omitted from the Australian version. The first, said to have been incorporated at the behest of the Japanese construction company Hazama Gumi, allegedly featured the construction of a one-mile (two-kilometer) high artificial mountain, approached by a three-mile (five-kilometer) runway on which a special linear motor vehicle would run to catapult a rocket into space. The second pointed out that Australia's influence in the Pacific region would grow because of its abundant energy reserves – a situation that made the country particularly interesting to resource-poor, strategically-dependent Japan (McCormack, 1990)
10 McCormack, 1991b: 39.
11 McCormack, 1991b: 39.
12 Hamilton, 1991: 32–8.
13 McCormack, 1991c: 154–5. Akita's location is 40°N, 140°E.
14 McCormack, 1991c: 159–60.
15 McCormack, 1991c: 161–2.
16 Hamilton, 1991: 17–19.
17 Tanaka, 1990: 73–7; McCormack, 1990: 133.
18 Hamilton, 1991: 54.
19 Rimmer, 1991: 112; Sasaki, 1991: 139–40.
20 Inkster, 1991: 28.
21 Sasaki, 1991: 139.
22 Rimmer, 1991: 113.
23 Inkster, 1991: 29.
24 Inkster, 1991: 80.
25 Inkster, 1991: 82.
26 Inkster, 1991: 107.
27 Morris-Suzuki, 1991: 131.
28 Inkster, 1991: 36–8.
29 Morris-Suzuki, 1991: 134–5.
30 Inkster, 1991: 44.
31 McCormack, 1990: 129.
32 Hamilton, 1991: 207.
33 Sasaki, 1991: 141.
34 Inkster, 1991: 43.
35 Hamilton, 1991: 62.
36 Hamilton, 1991: 120.
37 McCormack, 1991b: 46.

38 McCormack, 1991b: 47.
39 Inkster, 1991: 142.
40 Hamilton, 1991: 214.
41 Hamilton, 1991: 214.
42 Inkster, 1991: 48–9.
43 Inkster, 1991: 54.
44 Hamilton, 1991: 180.
45 McCormack, 1991b: 51–2.
46 Hamilton, 1991: 186; Inkster, 1991: 57.
47 Hamilton, 1991: 187, 190.
48 Hamilton, 1991: 131; Inkster, 1991: 57.
49 Hamilton, 1991: 195–6.
50 Inkster, 1991: 140.
51 Inkster, 1991: 45.
52 Chalkley and Winchester, 1991: 246.
53 MFP-Adelaide Management Board, 1991: n.p.
54 MFP-Adelaide Management Board, 1991: (1) 1.
55 MFP-Adelaide Management Board, 1991: (1) 2.
56 MFP-Adelaide Management Board, 1991: (7) 3.
57 MFP-Adelaide Management Board, 1991: (6) 1.
58 MFP-Adelaide Management Board, 1991: viii.
59 MFP-Adelaide Management Board, 1991: (2) 8.
60 McCormack, 1990: 135–6.
61 Morris-Suzuki, 1991: 123.
62 Sugimoto, 1990: 138.

CHAPTER 9

1 For a recent and sophisticated example, see Massey, Quintas, and Wield, 1991.
2 Marshall, 1920: 225.

CHAPTER 10

1 Hayek, 1944: 31.
2 Markusen, *et al.*, 1991.

REFERENCES

Andersson, Å.E. (1985a) "Creativity and Regional Development," *Papers of the Regional Science Association*, 56: 5–20.

—— (1985b) *Kreativitet: StorStadens Framtid*, Stockholm: Prisma.

Andersson, Å.E. and Strömquist, Ulf (1988) *K – Samhällets Framtid*, Stockholm: Prisma.

Ando, Y. (1983) "Kanagawa-ken Keizai no Tokucho nitsuite (Characteristics of Kanagawa Prefecture's Economy)," in Kanagawa-ken Kenmin-bu Kenshi Henshu-shitsu (ed.) *Kanagawa Kenshi Kakuron-hen 2 Sangyo Keizai*: 3–24, Yokohama: Kanagawa-ken Kosai-kai.

Anon (1989) "Un Pôle Européen à Massy," *Cahiers de l'Institut d'Aménagement et d'Urbanisme de la Région d'Ile-de-France*, 89: 33–49.

Anon (1990) *Proceedings of the Korea–UK International Symposium on High Tech Centers and Urban Development*, Taejon: South Korea.

Arthur, W.B. (1985) *Industry Location and the Economics of Agglomeration: Why a Silicon Valley*, Stanford, CA: Stanford University, Center for Economic Policy Research.

Arthur, W.B. (1986) *Industry Location Patterns and the Importance of Silicon Valley*, Stanford, CA: Stanford University, Food Research Institute.

Aydalot, P. (ed.) (1986a) *Milieux Innovateurs en Europe*, Paris: GREMI (privately printed).

—— (1986b) "Trajectoires technologiques et milieux innovateurs," in Aydalot, P. (ed.) *Milieux Innovateurs en Europe*: 345–61. Paris: GREMI (privately printed).

—— (1988) "Technological Trajectories and Regional Innovation in Europe," in Aydalot, P. and Keeble, D. (eds) *High Technology Industry and Innovative Environments: The European Experience*: 22–47, London: Routledge & Kegan Paul.

Aydalot, P., Keeble, D. (eds) (1988a) *High Technology Industry and Innovative Environments: The European Experience*, London: Routledge & Kegan Paul.

—— (1988b) "High-Technology Industry and Innovative Environments in Europe: An Overview," in Aydalot, P. and Keeble, D. (eds) *High Technology Industry and Innovative Environments: The European Experience*: 1–21, London: Routledge & Kegan Paul.

Azouaou M. and Magnaval R. (1986) *Silicon Valley: Un marché aux puces*, Paris: Fayard.

Bavaria (1985) *Microelectronics Industry in Bavaria*, Munich: Bavarian State Ministry of Economics.

—— (1987) *The Software Market in Bavaria*, Munich: Bavarian Ministry for Economic Affairs and Transport.

—— (1988) *Electronics in Bavaria*, Munich: Bavarian Ministry for Economic Affairs and Transport.
Bernstein, A., DeGrasse, B., Grossman, R., Paine, C. and Siegel, L. (1977) *Silicon Valley: Paradise or Paradox?: The Impact of High Technology Industry on Santa Clara County*, Mountain View, Calif.: Pacific Studies Center.
Bijker, W.E., Hughes T.P., and Pinch, T.J. (eds) (1987) *The Social Construction of Technological Systems*, Cambridge, MA: MIT Press.
Bloom, J.L. and Asano, S. (1981) "Tsukuba Science City: Japan Tries Planned Innovation," *Science*, 212: 1239–47.
Bluestone, B. and Harrison, B. (1982) *The Deindustrialization of America: Plant Closings, Community Abandonment, and the Dismantling of Basic Industry*, New York: Basic Books.
Borrus, M. (1989) *Competing for Control: America's Stake in Microelectronics*, Cambridge, MA: Ballinger.
Boudeville, J.-R. (1966) *Problems of Regional Economic Planning*, Edinburgh: Edinburgh University Press.
Broadbent, J. (1989) "'The Technopolis Strategy' versus Deindustrialization: High-Tech Development Sites in Japan," in Smith, M.P. (ed.) *Pacific Rim Cities in the World Economy (Comparative Urban and Community Research, Vol. 2)*: 231–53, New Brunswick: Transaction.
Brusco, S. (1982) "The Emilian Model: Productive Decentralisation and Social Integration," *Cambridge Journal of Economics*, 6: 167–84.
Camagni, R. and Rabellotti, R. (1986) "Innovation and Territory: The Milan High-Tech and Innovation Field," in Aydalot, P. (ed.) *Milieux Innovateurs en Europe*: 101–28, Paris: GREMI (privately printed).
Carey, C. and Gathright, M. (1985) "The Silicon Valley Worker," *San Jose Mercury News*, September (series of articles).
Carnoy, M., Castells, M., Cohen, S., Cardoso, F.H. (1993) *The New Global Economy in the Information Age*, University Park, PA: Pennsylvania State University Press.
Castells, M. (1967) *Les Politiques d'Implantation des entreprises industrielles dans la région parisienne*, Paris: Université de Paris, Thèse pour le Doctorat en Sociologie.
—— (1984) *Towards the Informational City?*, Berkeley: University of California, Institute of Urban and Regional Development, Working Paper 430, August.
—— (1989) *The Informational City*, Oxford: Basil Blackwell.
—— (1992) "Four Asian Tigers with a Dragon Head: State Intervention and Economic Development in the Asian Pacific Rim," in Appelbaum, R. and Henderson, J. (eds) *State and Society in the Pacific Rim*, Newbury Park and London: Sage.
Castells, M. and Hall, P. (eds) (1992) *Andalucía: Innovación Tecnológia y Desarrollo Económico*, Madrid: Espasa-Calpe.
Castells, M. and Tyson, L. (1988) "High Technology Choices Ahead: Restructuring Interdependence," in Sewell, J.W. and Tucker, S.K. (eds) *Growth, Exports, and Jobs in a Changing World Economy*, New Brunswick, NJ: Transaction.
Chalkley, B. and Winchester, H. (1991) "Green Light for the MFP," *Town and Country Planning*, 60: 244–6.
Christaller, W. (1966 (1933)) *Central Places in Southern Germany*, trans. C.W. Baskin, Englewood Cliffs, NJ: Prentice-Hall.
Cohen, S. and Zysman, J. (1987) *Manufacturing Matters: The Myth of the Postindustrial Economy*, New York: Basic Books.
Decoster, E. (1988) "La Création en Ile de France Sud," *Revue d'Economie Régionale et Urbaine*, 11: 489–517.
Decoster, E. and Tabaries, M. (1986) "L'Innovation dans un Pôle scientifique et

technologique, le cas de la cité scientifique Ile de France Sud," in Aydalot, P. (ed.) *Milieux Innovateurs en Europe*: 79–100, Paris: GREMI (privately printed).

Delius, F.C. (1977) *Unsere Siemens-Welt: Eine Festschrift zum 125jährigen Bestehen des Hauses S*, Berlin: Rotbuch Verlag.

Denison, E.W. (1985) *Productivity Growth in the U.S. Economy*, Washington, DC: The Brookings Institution.

Dorfman, N.S. (1982) *Massachusetts' High Technology Boom in Perspective: An Investigation of its Dimensions, Causes and of the Role of New Firms*, Cambridge, MA: MIT, Center for Policy Alternatives (CPA 82–2).

—— (1983) "Route 128: The Development of a Regional High Technology Economy," *Research Policy*, 12: 299–316.

Dosi, G., Freeman, C., Nelson, R., Silverberg, N., and Soete, L. (eds) (1988) *Technical Change and Economic Theory*, London: Pinter.

Douglass, M. (1988) "The Transnationalization of Urbanization in Japan," *International Journal of Urban and Regional Research*, 12 (3): 425–54.

Edgington, D.W. (1989) "New Strategies for Technology Development in Japanese Cities and Regions," *Town Planning Review*, 60: 1–27.

Fishman, K.D. (1981) *The Computer Establishmnent*, New York: McGraw Hill.

Florida, R. and Kenney, M. (1987) "Venture Capital, High Technology, and Regional Development," *Regional Studies*, 22: 33–48.

Forester, T. (1987) *High Tech Society: The Story of the Information Technology Revolution*, Oxford: Blackwell.

Freeman, C. (1987) *Technology Policy and Economic Performance: Lessons from Japan*, London: Frances Pinter.

Freiberger, P. and Swaine, M. (1984) *Fire in the Valley: The Making of the Personal Computer*, Berkeley: Osborne/McGraw Hill.

Fujita, K. (1988) "The Technopolis: High Technology and Regional Development in Japan," *International Journal of Urban and Regional Research*, 12: 556–94.

Fujitsu Corporation (1986) *Shashi III (Corporate History III)*, Tokyo: Fujitsu Corporation.

Fukyu Kouhou Senmon Iinkai, Tsukuba Kenkyu Gakuen Toshi Kenkyu Kikan Tou Renraku Kyougikai (1989) *Tsukuba Kenkyu Gakuen Toshi (Tsukuba Science City)*. Tsukuba: Tsukuba Shuppan-kai.

Furutate, H. (1982) "Technopolis Construction Concept: To Bring High Technology to Provincial Regions; Door Will Be Open to Foreign Firms, Too," *Journal of Japanese Trade & Industry*, 1: 10–13.

Gaffard, J.-L. (1986) "Restructuration de l'Espace économique et trajectoires technologiques," in Aydalot, P. (ed.) *Milieux Innovateurs en Europe*: 17–28, Paris: GREMI (privately printed).

Gakuen Toshi Mondai Kenkyu-Kai (1985) *Tsukuba Kenku Gakuen Toshi (Tsukuba Science City)*, Tokyo: Otsuki Shoten.

Giese, E. (1987) "The Demand for Innovation-Oriented Regional Policy in the Federal Republic of Germany: Origins, Aims, Policy Tools and Prospects of Realisation," in Brotchie, J., Hall, P., and Newton, P. (eds) *The Spatial Impact of Technological Change*: 240–53, London: Croom Helm.

Glasmeier, A. (1988) "The Japanese Technopolis Programme: High-Tech Development Strategy or Industrial Policy in Disguise?" *International Journal of Urban and Regional Research*, 12: 268–84.

Goetzeler, H. and Schoen, L. (1986) *Wilhelm und Carl Friedrich von Siemens: Die zweite Unternehmergeneration*, Stuttgart: Franz Steiner Verlag.

Goodman, R. (1979) *The Last Entrepreneurs*, New York: Basic Books.

Gordon, R. and Kimball, L. (1987) "The Impact of Industrial Structure on Global

High-Technology Industry," in Brotchie, J., Hall, P., and Newton, P. (eds) *The Spatial Impact of Technological Change*: 157–84, London: Croom Helm.

Gregory, G. (1985) "Science City: The Future Starts Here," *Far Eastern Economic Review*, 127: 43–8.

Grotz, R. (1989) "Technologische Erneuerung und Technologieorientierte Unternehmensgründungen in der Industrie in der Bundesrepublik Deutschland," *Geographische Rundschau*, 41: 266–72.

Guerrieri, P. (1991) *Technology and International Trade Performance in the Most Advanced Countries*, Berkeley: University of California, BRIE Working Paper 49, January.

Hägerstrand, T. (1967) *Innovation Diffusion as a Spatial Process*, Chicago: University of Chicago Press.

Hall, P. (1987) "The Geography of High-Technology Industry: An Anglo-American Comparison," in Brotchie, J., Hall, P., and Newton, P. (eds) *The Spatial Impact of Technological Change*: 141–56, London: Croom Helm.

—— (1988) "The Creation of the American Aerospace Complex, 1955–65: A Study in Industrial Inertia," in Breheny, M. (ed.) *Defence and Regional Development:* 102–21, London: Mansell.

—— (1990) *The Generation of Innovative Milieux: An Essay in Theoretical Synthesis*, Berkeley: University of California, Institute of Urban and Regional Development, Working Paper 505, March.

—— (1995 forthcoming) *Polis: Great Cities and their Golden Ages*, London: Harper Collins.

Hall, P., Breheny, M., McQuaid, R., and Hart, D. (1987) *Western Sunrise: The Genesis and Growth of Britain's Major High Technology Corridor*, London: Allen & Unwin.

Hall, P. and Preston, P. (1988) *The Carrier Wave: New Information Technology and the Geography of Innovation, 1846–2003*, London: Unwin Hyman.

Hamilton, W. (1991) *Serendipity City: Australia, Japan and the Multifunction Polis*, Sydney: Australian Broadcasting Corporation.

Hasegawa, M. (1983) "Denki Kikai Kogyo no Keisei to Hatten (Formation and Development of the Electric Machinery Industry)," in Kanagawa-ken Kenmin-bu Kenshi Henshu-shitsu (ed.) *Kanagawa Kenshi Kakuron-hen 2 Sangyo Keizai*: 261–304, Yokohama: Kanagawa-ken Kosai-kai.

Hayek, F.A. (1944) *The Road to Serfdom*, London: George Routledge.

Hepworth, M.E. (1989) *Geography of the Information Economy*, London: Belhaven Press.

Imai, K.-I. (1986) "Japan's Industrial Policy for High Technology Industry," in Patrick, H. (ed.) *Japan's High Technology Industries: Lessons and Limitations of Industrial Policy*: 137–69, Seattle: University of Washington Press.

Industrial Structure Council (1980) *The Vision of MITI Policies in 1980s*, Tokyo: Ministry of International Trade and Industry.

Inkster, I. (1991) *The Clever City: Japan, Australia and the Multifunction Polis*, Sydney: Sydney University Press.

Itakura, K. and Takeuchi, A. (1980) "Keihin Region," in Murata, K. and Ota, I. (eds) *An Industrial Geography of Japan*: 47–65, New York: St Martin's Press.

Jacobs, J. (1984) *Cities and the Wealth of Nations: Principles of Economic Life*, New York: Random House.

Japan Economic Survey (1984) "Commentary: Governor Morihiko Hiramatsu on Regional Investment in Japan," *Japan Economic Survey*, 8: 6–11.

Johansson, B. (1987) "Information Technology and the Viability of Spatial Networks," *Papers of the Regional Science Association*, 61: 51–64.

Johansson, B. and Westin, L. (1987) "Technical Change, Location, and Trade," *Papers of the Regional Science Association*, 62: 13–25.

Johnson, C.A. (1982) *MITI and the Japanese Miracle: the Growth of Industrial Policy, 1925–1975*, Stanford, CA: Stanford University Press.

Johnstone, B. (1988) "Watermelons, Mushrooms . . . and Chips," *Far Eastern Economic Review*, 140: 78–83.

Kanagawa-ken Kikaku-bu Tokei-ka (1988) *Chiiki Keizai no Seicho to Kozo Henka*: 367–82, ed. Keizai Kikaku-cho Keizai Kenkyu-jo Kokumin Shotoku-bu, Tokyo: Okura-sho Insatsu-kyoku.

Kanagawa Prefectural Government (1985) *The History of Kanagawa*, Yokohama: Kanagawa Prefectural Government.

Kawashima, T. and Stöhr, W. (1988) "Decentralized Technology Policy: The Case of Japan," *Environment and Planning C: Government and Policy*, 6: 427–39.

Kawashima, T. and Taketoshi, M. (1987) "Regional Economic Policies: Through Development of High Technology Oriented Industries," *Gakushuin Economic Papers*, 23: 13–37.

Keeble, D.E. (1968) "Industrial Decentralization and the Metropolis: The North-West London Case," *Transactions and Papers, Institute of British Geographers*, 44: 1–54.

—— (1988) "High-Technology Industry and Local Environments in the United Kingdom," in Aydalot, P. and Keeble, D. (eds) *High Technology Industry and Innovative Environments: The European Experience*: 65–98, London: Routledge & Kegan Paul.

—— (1989) "High-Technology Industry and Regional Development in Britain: The Case of the Cambridge Phenomenon," *Environment and Planning C; Government and Policy*, 7: 153–72.

Keller, J. (1981) "The Production Worker in Electronics: Industrialization and Labour Development in California's Santa Clara Valley," Ann Arbor: University of Michigan, Ph.D. Dissertation, unpublished.

Klingbeil, D. (1987a) "Münchens Wirtschafts- und Bevölkerungsentwicklung nach dem II. Weltkrieg," in Geipel, R., Hartke, W., and Heinritz, G. (eds) *München: Ein Sozialgeographische Exkursionsführer (Münchener Geographische Hefte Nr. 55/56)*: 43–66, Kallmünz/Regensburg: Michael Lassleben.

—— (1987b) "Epochen der Stadtgeschichte und der Stadtstrukturentwicklung," in Geipel, R., Hartke, W., and Heinritz, G. (eds) *München: Ein Sozialgeographische Exkursionsführer (Münchener Geographische Hefte Nr. 55/56)*: 67–100, Kallmünz/Regensburg: Michael Lassleben.

—— (1987c) "Grundzüge der Stadtstrukturellen Entwicklung nach der II. Weltkrieg," in Geipel, R., Hartke, W., and Heinritz, G. (eds) *München: ein Sozialgeographische Exkursionsführer (Münchener Geographische Hefte Nr. 55/56)*: 101–40, Kallmünz/Regensburg: Michael Lassleben.

Kobayashi, M. (1983) "Keihin Kogyo Chitai niokeru Chusho Kogyo (Small- and Medium-Sized Manufacturing in Keihin Industrial Belt)," in Kanagawa-ken Kenmin-bu Kenshi Henshu-shitsu (ed.) *Kanagawa Kenshi Kakuron-hen 2 Sangyo Keizai*: 119–61, Yokohama: Kanagawa-ken Kosai-kai.

Koestler, A. (1975) *The Act of Creation*, London: Picador.

Kondratieff, N.D. (1935) "The Long Waves in Economic Life," *The Review of Economic Statistics*, 17: 105–15.

—— (1984) *The Long Wave Cycle*, New York: Richardson & Snyder.

Kranzberg, M. and Pursell, C.W. (eds) (1967) *Technology in Western Civilization*, New York: Oxford University Press.

Kreis der Freunde Alt-Münchens (1957) *München im Wandel der Jahrhunderte: Bilder aus der Sammlung Proebst*, München: F. Bruckmann.

Kroll, C. and Kimball, L. (1986) *The Santa Clara Valley R and D Dilemma: The Real Estate Industry and High Tech Growth*, Berkeley: University of California, Center for Real Estate and Urban Economics, Working Paper.

Lafitte, P. (1985) "Sophia Antipolis and its Impact on the Côte d'Azur," in Gibb, J.M. (ed.) *Science Parks and Innovation Centres: Their Economic and Social Impact*, New York: Elsevier.

Lampe, D.R. (1984) "Das M.I.T. und die Entwicklung der Region Boston," in Schwarz, K. (ed.) *Die Zukunft der Metropolen: Paris, London, New York, Berlin: Aufsätze*: 554–9, Berlin: Technische Universität.

—— (ed.) (1988) *The Massachusetts Miracle: High Technology and Economic Revitalization*, Cambridge, MA: MIT Press.

Lee, C.O. (1988) *Science and Technology Policy of Korea*, Seoul: Korea Advanced Institute of Science and Technology.

Lim, H.-C. (1985) *Dependent Development in Korea 1963–79*, Seoul: Seoul National University Press.

Loria, J. (1984) "Das Massachusetts Institute of Technology und die Entwicklung der Region Boston," in Schwarz, K. (ed.) *Die Zukunft der Metropolen: Paris, London, New York, Berlin: Katlog zur Ausstellung*: 128–46, Berlin: Technische Universität.

Lösch, A. (1954 (1944)) *The Economics of Location*, trans. W.H. Woglom, New Haven: Yale University Press.

McCormack, G. (1990) "And Shall the Multifunction Polis yet be Built?" in James, P. (ed.) *Technocratic Dreaming: Of Very Fast Trains and Japanese Designer Cities*: 129–37, Sydney: Left Book Club.

—— (ed.) (1991a) *Bonsai Australia Banzai: Multifunction Polis and the Making of a Special Relationship with Japan*, Leichhardt, NSW: Pluto Press Australia.

—— (1991b) "Coping with Japan: The MFP Proposal and the Australian Response," in McCormack, G. (ed.) *Bonsai Australia Banzai: Multifunction Polis and the Making of a Special Relationship with Japan*, 34–67. Leichhardt, NSW: Pluto Press Australia.

—— (1991c) "'High-Touch': Leisure and Resorts in Japanese Development," in McCormack, G. (ed.) (1991) *Bonsai Australia Banzai: Multifunction Polis and the Making of a Special Relationship with Japan*: 152–69, Leichhardt, NSW: Pluto Press Australia.

Malone, M.S. (1985) *The Big Score: The Billion-Dollar Story of Silicon Valley*, Garden City, NY: Doubleday.

Mandel, E. (1973) *Late Capitalism*, Atlantic Highlands, NJ: Humanities Press.

—— (1980) *Long Waves of Capitalist Development*, Cambridge: Cambridge University Press.

Markusen, A. (1985) *Profit Cycles, Oligopoly, and Regional Development*, Cambridge, MA: MIT Press.

—— (1987) *Regions: The Economics and Politics of Territory*, Totowa: Rowman & Littlefield.

Markusen, A., Hall, P., Campbell, S., and Deitrick, S. (1991) *The Rise of the Gunbelt: The Military Remapping of Industrial America*, New York: Oxford University Press.

Markusen, A., Hall, P., and Glasmeier, A. (1986) *High-Tech America: The What, How, Where and Why of the Sunrise Industries*, Boston: George Allen & Unwin.

Marshall, A. (1920 (1890)) *Principles of Economics*, London: Macmillan.

Martin, J.E. (1966) *Greater London: An Industrial Geography*, London: Bell.

Maruyama, M. (1985) "Report on a new Technological Community: The Making of Technopolis in an International Context," *Technological Forecasting and Social Change*, 27: 75–98.

Masser, I. (1990) "Technology and Regional Development Policy: A Review of Japan's Technopolis Programme," *Regional Studies*, 24: 41–53.

Massey, D. and Meegan, R. (1982) *The Anatomy of Job Loss: The How, Why and Where of Employment Decline*, London: Methuen.

Massey, D., Quintas, P., and Wield, D. (1991) *High-Tech Fantasies: Science Parks in Society, Science and Space*, London: Routledge.

MFP-Adelaide Management Board (1991) *Report on the Feasibility of MFP-Adelaide*, prepared for the Commonwealth and South Australian Governments, Adelaide: MFP-Adelaide Management Board.

MITI (Ministry of International Trade and Industry, Industrial Location and Environmental Protection Bureau) (1988a) *Zunou Ritchihou no Kaisetsu: Sangyo no Zunou Bubun no Chiiki Shuseki ni Mukete (Commentary on the Brain Location Law: Toward Regional Agglomeration of the Brain Function of Industries)*, Tokyo: Tsusho Sangyo Chosakai.

—— (1988b) "Tekunoporis no Kensetsu Joukyo (Progress in the Technopolis Programme)," *Sangyo Ritchi*, (2) 27: 46–52.

MITI (Ministry of International Trade and Industry) (1990) *Outline and Present Status of the Technopolis Project*, Tokyo: MITI, Industrial Location and Environmental Protection Bureau (mimeo).

Miyakawa, Y. (1980) "The Location of Modern Industry in Japan," in Association of Japanese Geographers (ed.) *Geography of Japan*: 265–98, Tokyo: Teikoku-Shoin.

Mollenkopf, J.H. (1983) *The Contested City*, Princeton: Princeton University Press.

Monck, C.S.P., Porter, R.B., Quintas, P., Storey, D.J., and Wynarczyk, P. (1988) *Science Parks and the Growth of High Technology Firms*, London: Croom Helm.

Monk, P. (1989) *Technological Change in the Information Economy*, London: Pinter.

Morris-Suzuki, T. (1990) "From Technopolis to Multifunction Polis: High-Tech Cities in Japan and Australia," in Mouer, R. and Sugimoto, Y. (eds) *The MFP Debate: A Background Reader*: 67–78, Melbourne: LaTrobe University Press.

—— (1991) "The MFP and the Japanese Development Model," in McCormack, G. (ed.) *Bonsai Australia Banzai: Multifunction Polis and the Making of a Special Relationship with Japan*: 123–137, Leichhardt, NSW: Pluto Press Australia.

Mouer, R. and Sugimoto, Y. (eds) (1990) *The MFP Debate: A Background Reader*, Melbourne: LaTrobe University Press.

Murata, K. (1988) "The Technopolis: Concept and Present Situation," *Journal of Economics (Keizaigaku Ronsan), Chuo University*, 29/5: 39–51.

Murata, K. and Takeuchi, A. (1987) "The Regional Division of Labour: Machinery Manufacturing, Microelectronics and R&D in Japan," in Hamilton, F.E.I. (ed.) *Industrial Change in Advanced Economies*: 213–39, London: Croom Helm.

Mutlu, S. (1979) *Inter-regional and International Mobility of Industrial Capital: The Case of American Automobile and Electronics Industries*, Berkeley: University of California, Ph.D. Dissertation in City and Regional Planning, unpublished.

Nakano, T. (1988) "Tekunoporis Kensetsu wa Aratana Sousouki e (The Technopolis Programme Enters a New Period)," *Sangyo Ritchi*, 27 (6): 37–40.

Nelson, R.E. (1981) "Research on Productivity Growth and Productivity Differences: Dead Ends and New Departures," *Journal of Economic Literature*, 19: 1029–64.

Nihon Denki Kabushiki Kaisha Shashi Hensanshitsu (1972) *Nihon Denki Kabushiki Kaisha Nanajunen-shi (NEC's Seventy-Year History)*, Tokyo: Nihon Denki Kabushiki Kaisha.

Nishioka, H. (1985) "High Technology Industry: Location, Regional Development and International Trade Frictions," *Aoyama Keizai Ronshu*, 36 (2, 3, and 4): 295–341.

Nöhbauer, H.F. (1982) *München: Eine Geschichte der Stadt und ihrer Bürger*, München: Süddeutscher Verlag.
Oakey, R. (1984) *High Technology Small Firms: Regional Development in Britain and the United States*, London: Pinter.
Ohmae, K. (1990) *The Borderless World: Power and Strategy in the Interlinked Economy*, New York: Harper & Row.
Okimoto, D.I. (1989) *Between MITI and the Market: Japanese Industrial Policy for High Technology*, Stanford, CA: Stanford University Press.
Onda, M. (1988) "Tsukuba Science City Complex and Japanese Strategy," in Smilor, R.W., Kozmetsky, G., and Gibson, D.V. (eds) *Creating the Technopolis: Linking Technology Commercialization and Economic Development*: 51–68, Cambridge, MA: Ballinger Publishing Company.
Pallot, J. and Shaw, D.J.B. (1981) *Planning in the Soviet Union*, London: Croom Helm.
Park, Y. (1986) "Manufacturing Decentralization and Regional Productivity Change," Berkeley: University of California, Ph.D. Dissertation in City and Regional Planning, unpublished.
Pavitt, K. (1988) "International Patterns of Technological Accumulation," in Hood, N. and Vahlne, J.E. (eds) *Strategies in Global Competition*, London: Croom Helm.
Perrin, J.-C. (1986a) "La Phénomène Sophia-Antipolis dans son environnement régional," in Aydalot, P. (ed.) *Milieux Innovateurs en Europe*: 283–302, Paris: GREMI (privately printed).
—— (1986b) "Les P.M.E. de HT à Valbonne Sophia Antipolis," *Révue d'Economie Régionale et Urbaine*, 9: 629–43.
—— (1988) "New Technologies, Local Synergies and Regional Policies in Europe," in Aydalot, P. and Keeble, D. (eds) *High Technology Industry and Innovative Environments: The European Experience*: 139–62, London: Routledge & Kegan Paul.
Perroux, F. (1961) *L'Economie du XX siècle*, Paris: Presses Universitaires de France.
—— (1965) *La Pensée économique de Joseph Schumpeter: les dynamiques du capitalisme*, Genève: Droz.
Piore, M.J. and Sabel, C.F. (1984) *The Second Industrial Divide: Possibilities for Prosperity*, New York: Basic Books.
Pirenne, H. (1914) "Les Périodes de l'histoire sociale du capitalisme," *Bulletin de l'Academie Royale de Belgique*, 5: 258–99.
Porat, M.U. (1977) *The Information Economy: Definition and Measurement*, Washington, DC, US Department of Commerce, Office of Telecommunications, OT Special Publication 77–12 (1), May.
Pred, A. (1977) *City-Systems in Advanced Economies: Past Growth, Present Processes and Future Development Options*, London: Hutchinson.
Quéré, M. (1990) *Sophia-Antipolis dans le Contexte Français*, Paris: GIP "Mutations Industrielles".
Ricardo, D. (1973 (1817)) *The Principles of Political Economy and Taxation*, London: Dent.
Richardson, H.W. and Hwang, M.-C. (eds) (1988) *Urban and Regional Policy in Korea and International Experiences*, Seoul: KRIHS.
Rimmer, P. (1991) "Global Techno-Belts in the Twenty-first Century," in McCormack, G. (ed.) *Bonsai Australia Banzai: Multifunction Polis and the Making of a Special Relationship with Japan*: 103–22. Leichhardt, NSW: Pluto Press Australia.
Roe, J.W. (1916) *English and American Tool Makers*, New Haven: Yale University Press.
Rogers, E.M. and Larsen, J.K. (1984) *Silicon Valley Fever: Growth of High-Technology Culture*, New York: Basic Books.

Roper-Dennis, N.E. (1984) "MITI's Technopolis Plan: A New Brand of Regional Development," *Oriental Economist*, 52: 56–8.
Rosenberg, N. (1976) *Perspectives on Technology*, Cambridge: Cambridge University Press.
Sakamoto, K. (1988) "Jouhou Shori Sangyo no Chiikiteki Tenkai: Tekunoporis Ken-iki tono Kanren de (Spatial Distribution of the Information Processing Industry: In Relation to Technopolis Prefectures)," in Tasuku N. (ed.) *Sentan Gijutsu to Tekunoporis*: 58–78, Tokyo: Nihon Keizai Hyoronsha.
Sasaki, M. (1991) "Japan, Australia and the Multifunctionpolis," in McCormack, G. (ed.) *Bonsai Australia Banzai: Multifunction Polis and the Making of a Special Relationship with Japan*: 138–51, Leichhardt, NSW: Pluto Press Australia.
Sase, M. (1982) "Tekunoporis Oitachi no Ki (A Story of the Birth and Childhood of the Technopolis Programme)," *Sangyo Ritchi*, 21(9): 6–11.
Saucier, P. (1987) *Specialisation Internationale et competitivité de l'économie japonaise*, Paris: Economica.
Saxenian, A.L. (1981) *Silicon Chips and Spatial Structure: The Industrial Basis of Urbanization in Santa Clara Valley, California*, Berkeley: University of California, Institute of Regional Development, Working Paper 345.
Saxenian, A.L. (1985) "Silicon Valley and Route 128: Regional Prototypes or Historic Exceptions?" in Castells, M. (ed.) *High Technology, Space and Society*: 81–105, Beverly Hills and London: Sage.
—— (1988) "The Cheshire Cat's Grin: Innovation and Regional Development in England," *Technology Review*, 91/2: 67–75.
—— (1989a) *The Political Economy of Industrial Adaptation in Silicon Valley*, Cambridge, MA: Massachusetts Institute of Technology, Department of Political Science, Ph.D. Dissertation, June, unpublished.
—— (1989b) "The Cheshire Cat's Grin: Innovation, Regional Development and the Cambridge Case," *Economy and Society*, 18: 448–77.
—— (1990) "Regional Networks and the Resurgence of Silicon Valley," *California Management Review*, 33: 89–112.
—— (1991) "The Origins and Dynamics of Production Networks in Silicon Valley," *Research Policy*, 20: 423–37.
—— (1992) "Contrasting Patterns of Business Organization in Silicon Valley," *Environment and Planning D, Society & Space*, 10: 377–91.
—— (1993) *Regional Networks: Industrial Adaptation in Silicon Valley and Route 128*, Cambridge, MA: Harvard University Press.
Schiffer, J. (1989) *Soviet Regional Economic Policy*, London: Macmillan.
Schreyer, K. (1969) *Bayern, ein Industriestaat: Die importierte Industrialisierung. Das wirtschaftliche Wachstum nach 1945 als Ordnungs- und Strukturproblem*, Wien: Olzog.
Schumpeter, J.A. (1939, 1982) *Business Cycles: A Theoretical, Historical and Statistical Account of the Capitalist Process* (2 vols), New York and London: McGraw Hill. Reprinted Philadelphia: Porcupine Press.
—— (1942) *Capitalism, Socialism and Democracy*, New York: Harper & Row.
Scott, A.J. (1982) "Locational Patterns and Dynamics of Industrial Activity in the Modern Metropolis," *Urban Studies*, 19: 114–42.
—— (1983a) "Industrial Organization and the Logic of intra-Metropolitan Location: I. Theoretical Considerations," *Economic Geography*, 59: 233–50.
——(1983b) "Industrial Organization and the Logic of intra-Metropolitan Location: II. A Case Study of the Printed Circuits Industry in the Greater Los Angeles Region," *Economic Geography*, 59: 343–67.
—— (1984a) "Industrial Organization and the Logic of intra-Metropolitan Location: III. A Case Study of the Women's Dress Industry in the Greater Los Angeles Region," *Economic Geography*, 60: 3–27.
—— (1984b) "Territorial Reproduction and Transformation in a Local Labor

Market: The Animated Film Workers of Los Angeles," *Environment and Planning D: Society and Space*, 2: 277–307.
—— (1985) "Industrialization and Urbanization: A Geographical Agenda," *Annals of the Association of American Geographers*, 76: 25–37.
—— (1986a) "Location Processes, Urbanization and Territorial Development: An Exploratory Essay," *Environment and Planning A*, 17: 479–501.
—— (1986b) "High Technology Industry and Territorial Development: The Rise of the Orange County Complex, 1955–1984," *Urban Geography*, 7: 3–45.
—— (1986c) "Industrial Organization and Location: Division of Labor, The Firm and Spatial Process," *Economic Geography*, 62: 215–31.
—— (1987) "The Semiconductor Industry in South-East Asia: Organization, Location, and the International Division of Labour," *Regional Studies*, 21: 143–60.
—— (1989a) *New Industrial Spaces*, London: Pion.
—— (1989b) *The Technopoles of Southern California* (UCLA Research Papers in Economic and Urban Geography, 1), Los Angeles: UCLA, Department of Geography.
Scott, A.J. and Angel, D.P. (1987) "The U.S. Semiconductor Industry: A Locational Analysis," *Environment and Planning, A*, 19: 875–912.
Scott, A.J. and Mattingly, D. (1989) "The Aircraft and Parts Industry in Southern California: Continuity and Change from the Inter-War Years to the 1990s," *Economic Geography*, 65: 48–71.
Scott, A.J. and Paul, S. (1990) "Collective Order and Economic Coordination in Industrial Agglomerations: The Technopoles of Southern California," *Environment and Planning, C: Government and Policy*, 8: 179–93.
Scott, A.J. and Storper, M. (1986a) *Production, Work, Territory: The Geographical Anatomy of Industrial Capitalism*, Boston: Allen & Unwin.
—— (1986b) "Industrial Change and Territorial Organization: A Summing Up," in Scott, A.J. and Storper, M., *Production, Work, Territory: The Geographical Anatomy of Industrial Capitalism*: 301–11, Boston: Allen & Unwin.
Segal Quince Wicksteed (1985) *The Cambridge Phenomenon: The Growth of High Technology Industry in a University Town*, Cambridge: Segal Quince Wicksteed.
—— (1988) *Universities, Enterprises and Local Economoic Development: An Exploration of Links* (A Report for the Manpower Services Commission), London: HMSO.
Siegel, L. and Markoff, J. (1985) *The High Cost of High Tech: The Dark Side of the Chip*, New York: Harper & Row.
Siemens, G. (1957a) *History of the House of Siemens*, Vol. I: *the Era of Free Enterprise*, Freiburg and Munich: Karl Alber.
—— (1957b) *History of the House of Siemens*, Vol. II: *The Era of World Wars*, Freiburg and Munich: Karl Alber.
Smith, D.H. (1933) *The Industries of Greater London: Being a Survey of the Recent Industrialisation of the Northern and Western Sectors of Greater London*, London: P.S. King.
Soete, L. (1991) *Technology and Economy in a Changing World* (Background Paper, OECD Conference on Technology and the Global Economy), Paris: OECD.
Soja, E. (1986) "Taking Los Angeles Apart: Some Fragments of a Critical Human Geography," *Environment and Planning, D: Society and Space*, 4, 255–72.
Solow, R. (1957) "Technical Changes and the Aggregate Production Function," *Review of Economics and Statistics*, 39: 312–20.
Sophia-Antipolis (1989) *Sophia-Antipolis: Le Site Intelligent de l'Europe*, Sophia-Antipolis: Société Anonyme Sophia-Antipolis.
Stöhr, W.B. (1986) "Territorial Innovation Complexes," in Aydalot, P. (ed.) *Milieux Innovateurs en Europe*: 29–54, Paris: GREMI (privately printed).
Storper, M. and Scott, A.J. (1986) "Production, Work, Territory: Contemporary Realities and Theoretical Tasks," in Scott, A.J. and Storper, M., *Production, Work,*

Territory: The Geographical Anatomy of Industrial Capitalism: 3–15, Boston: Allen & Unwin.

Sugimoto, Y. (1990) "High-Tech Cities for Lonely Technocrats," in James, P. (ed.) *Technocratic Dreaming: Of Very Fast Trains and Japanese Designer Cities*: 138–145, Sydney: Left Book Club.

Takahashi, N. (1981) "A New Concept in Building: Tsukuba Academic New Town," *Ekistics*, 48: 302–6.

Takanashi, M. and Hyodo, T. (1963) "Kogyo (Industry)," in Okochi, K. (ed.) *Keihin Kogyo Chitai no Sangyo Kozo (Industrial Structure of Keihin Industrial Belt)*: 49–116, Tokyo: Tokyo University Press.

Tanaka, Y. (1990) "The Japanese Political-Industrial Complex," in James, P. (ed.) *Technocratic Dreaming: Of Very Fast Trains and Japanese Designer Cities*: 71–7. Sydney: Left Book Club.

Tatsuno, S.M. (1986) *The Technopolis Strategy: Japan, High Technology, and the Control of the Twenty-first Century*, New York: Prentice Hall.

—— (1988) "Building a Japanese Technostate: MITI's Technopolis Program," in Smilor, R.W., Kozmetsky, G., and Gibson, D.V. (eds) *Creating the Technopolis: Linking Technology, Commercialization and Economic Development*: 3–21, Cambridge, MA: Ballinger Publishing Company.

—— (1990) *Created in Japan: From Imitators to World-Class Innovators*, New York: Harper & Row.

Thürauf, G. (1973) "Über einige Aspekte de Entwicklung der Münchener Industrie," *Mitteilungen der Geographischen Gesellschaft in München*, 58: 5–18.

—— (1975) *Industriestandorte in der Region München: Geographische Aspekte des Wandels industrieller Strukturen. (Münchener Studien zur Sozial- und Wirtschaftsgeographie, Band 16)*, Kallmünz/Regensburg: Michael Lassleben.

Toda, T. (1987) "The Location of High-Technology Industry and the Technopolis Plan in Japan," in Brotchie, J., Hall, P., and Newton, P. (eds) *The Spatial Impact of Technological Change*: 271–83, London: Croom Helm.

Tokyo Shibaura Electric Corporation (1977) *Toshiba Hyakunen-shi (Toshiba's Hundred-Year History)*, Kawasaki: Tokyo Shibaura Electric Corporation.

Vance, J.E. (1970) *The Merchant's World: The Geography of Wholesaling*, Englewood Cliffs, NJ: Prentice-Hall.

Vernon, R. (1966) "International Investment and International Trade in the Product Cycle," *Quarterly Journal of Economics*, 80: 190–207.

von Thünen, J.H. (1966 (1826)) *von Thünen's Isolated State*, ed. P. Hall, trans. C.M. Wartenberg, Oxford: Pergamon.

Weber, A. (1929 (1909)) *Alfred Weber's Theory of the Location of Industries*, trans. C.J. Friedrich, Chicago: Chicago University Press.

Weiher, S. von (1987) *Berlins Weg zur Elektropolis: Ein Beitrag zur Technik- und Industriegeschichte an der Spree*, supplementary chapter by G. Vetter, 2nd edn, Göttingen and Zurich: Muster-Schmidt Verlag.

Weiher, S. von and Goetzeler, H. (1981) *Weg und Wirken der Siemens-Werke im Fortschritt der Elektrotechnik 1847–1980: Ein Beitrag zur Geschichte der Elektroindustrie (Zeitschrift fur Unternehmensgeschichte, Beiheft 21)*, Wiesbaden: Franz Steiner.

Wildes, K.L. and Lindgren, N.A. (1985) *A Century of Electrical Engineering and Computer Science at M.I.T., 1882–1982*, Cambridge, MA: MIT Press.

Wood, A. and Joyce, W. (eds) (1991) *Modern Siberia: Social and Economic Development*, London: Routledge.

Yamasaki, A. (1990) "Tekunoporis Fibah no Aenai Ketsumatsu (Disappointing Outcome of Technopolis Fever)," *Ekonomisuto*, 68/12: 28–33.

INDEX